TWENTIETH CENTURY INTERPRETATIONS
OF

BOSWELL'S
LIFE OF
JOHNSON

A Collection of Critical Essays
Edited by
JAMES L. CLIFFORD

Prentice-Hall, Inc. A SPECTRUM BOOK *Englewood Cliffs, N. J.*

Passages from Boswell's original notes and journals are included by permission of Yale University and the McGraw-Hill Book Company.

PRENTICE-HALL INTERNATIONAL, INC. (*London*)
PRENTICE-HALL OF AUSTRALIA, PTY. LTD. (*Sydney*)
PRENTICE-HALL OF CANADA, LTD. (*Toronto*)
PRENTICE-HALL OF INDIA PRIVATE LIMITED (*New Delhi*)
PRENTICE-HALL OF JAPAN, INC. (*Tokyo*)

Contents

PART THREE—*View Points*

Introduction

by James L. Clifford

Boswell's *Life of Johnson* has been universally acknowledged as one of the great books of the Western world. Yet only in the last half century have we had any clear idea of the nature of Boswell's achievement, or of the way in which he worked. Generally accepted for most of the nineteenth century was Macaulay's theory that the *Life* was an accidental masterpiece, produced by a fool who happened to have a perfect subject. Boswell's success, it was assumed, merely lay in his skill as a reporter. No one ever thought of him as a great writer or as a major creative artist. Instead critics and scholars tended to annotate the *Life* extensively, while using it as a central text to build up a picture of the colorful Johnson circle. The clubbable Johnson and his cronies, with their witty talk and eccentric behavior, were stressed, rather than the greatness of the work as an artistic biography.

In the 1920s, following the fantastic discovery of a portion of Boswell's archives at Malahide Castle outside Dublin, and the subsequent further discoveries there and in Scotland, the emphasis gradually began to change.[1] For one thing, Boswell's reputation as a revealing diarist steadily mounted. His London journal of 1762–63, discovered at Fettercairn House in Scotland in 1930, but not published until twenty years later, became a best seller and was widely hailed as one of the most fascinating autobiographical documents in all literature. Thus today, irrespective of his connection with Johnson, Boswell stands out as one of the best of subjective chroniclers.

In addition to changing Boswell's overall reputation as a creative writer, the new discoveries provide extensive evidence concerning Boswell's method of recording conversations, the question of his basic accuracy, and his particular technique in combining all his diverse material into a readable biography. Now we are able to evaluate more

[1] See, for example, F. A. Pottle, "History of the Boswell Papers," *Boswell's London Journal*, limited edition (London: William Heinemann Ltd., 1951), pp. xi–xlii, and "Boswell Revalued" in *Literary Views*, ed. Carroll Camden (Chicago: Univ. of Chicago Press, 1964), pp. 79–91. Also John N. Morris, *Versions of the Self* (New York: Basic Books, 1966), pp. 171–210, etc.

clearly the importance of the *Life* in the development of modern biography, as a key text in the setting up of the modern ethical problem of how much a biographer can tell.

Boswell and Johnson

When Boswell met Samuel Johnson on May 16, 1763, he was in his twenty-third year. Born in Edinburgh October 29, 1740, he was the son of Alexander Boswell, Laird of Auchinleck, one of the important Lords of Sessions.[2] He had been well educated, at first by private tutor, later attending the University of Edinburgh for six years, and for a time the lectures of Adam Smith in Glasgow. From the start his father had been determined that he study law, though his own inclination was more for a commission in the Guards. An avid theatre goer and occasional poet, he found the ladies—all kinds—a source of irresistible attraction. He had run away to London in the spring of 1760, when he stayed for several months, and in the early fall of 1762 came again for a longer visit, still vainly hoping for an army commission. By this time a devoted keeper of a journal, he found London full of delights and disappointments.

The famous man he met in Tom Davies' back parlor in May 1763— the pensioned lexicographer and essayist with the greater part of his career behind him—was thirty years his senior. Inevitably, from the start, the relationship was that of a youthful admirer to an eccentric middle-aged celebrity. But it soon became much more than that, for the youthful Scot, for all his levity and rakish propensities, had a serious side. This Johnson soon found out, as he grew to respect the keen intelligence behind Boswell's gay exterior. Through Boswell's revealing accounts it is fascinating to watch the development of their close friendship.

Nevertheless, it would be a mistake to assume, as have so many in the past, that Johnson immediately became the center of Boswell's existence. Because he later gathered together in one place so many intimate descriptions of their meetings, Boswell does tend to give that impression. Actually, Johnson was only one of a number of important figures who move through Boswell's journals. Indeed, the great portion of Boswell's life was lived outside the direct influence of Johnson. After the first period of association in 1763, Boswell was away on the Con-

---

[2] See F. A. Pottle, *James Boswell: the Earlier Years 1740–1769* (New York: McGraw-Hill Book Co., 1966). The second volume of this life is being prepared in collaboration with Frank Brady. An older short biography is that by W. K. Leask (Edinburgh, 1897). Popular lives by C. E. Vulliamy (London: Geoffrey Bles, 1932) and D. B. Wyndham Lewis (London: Eyre & Spottiswoode, 1946) are not to be wholly trusted.

tinent for over two and a half years, and after his return his home was in Scotland. There he was an active lawyer, with many family responsibilities. He would see Johnson on his trips to London, and, of course, they kept in touch by occasional letters. Sometimes there were long periods when they were apparently out of touch. It has been estimated that they could have been together not more than about 425 days, of which 100 would have been during the Hebridean tour in 1773.[3] Even so, it would also be a mistake to play down too much the influence of the great moralist on the younger man. He did become a kind of father figure for Boswell.[4]

It is not completely certain when Boswell first considered writing the life of Johnson. He may have had vague thoughts about the possibility of such an undertaking from the start, but there are no definite references to it until later. In May 1768 he asked Johnson's permission to publish his letters at a later time,[5] and he wrote in his journal for March 31, 1772, "I have a constant plan to write the life of Mr. Johnson. I have not told him of it yet, nor do I know if I should tell him."[6]

When in the Hebrides the next year, Boswell recorded on October 14:

> The Sunday evening that we sat by ourselves at Aberdeen, I asked him several particulars of his life from his early years, which he readily told me, and I marked down before him. This day I proceeded in my inquiries, also marking before him. I have them on separate leaves of paper. I shall lay up authentic materials for THE LIFE OF SAMUEL JOHNSON, LL.D., and if I survive him, I shall be one who shall most faithfully do honour to his memory. I have now a vast treasure of his conversation at different times since the year 1762 [1763] when I first obtained his acquaintance; and by assiduous inquiry I can make up for not knowing him sooner.[7]

That Johnson approved is evident from a later note which Boswell added: "It is no small satisfaction to me to reflect that Dr. Johnson read this, and, after being apprised of my intention, communicated to me, at subsequent periods, many particulars of his life, which probably could not otherwise have been preserved."[8] But it must be admitted

---

[3] P. A. W. Collins, "Boswell's Contact with Johnson," *Notes & Queries*, April 1956, pp. 163–66.

[4] See, for example, Ian Ross, "Boswell in Search of a Father? or a Subject?" *Review of English Literature*, V (January 1964), 19–34, and other psychological studies of Boswell.

[5] *Life*, II, 60.

[6] *Boswell for the Defence*, ed. W. K. Wimsatt, Jr. and F. A. Pottle (New York: McGraw-Hill Book Co., 1959), p. 83.

[7] *Boswell's Journal of a Tour to the Hebrides*, ed. F. A. Pottle and C. H. Bennett (New York: Viking Press, 1936), p. 300.

[8] *Ibid.*, note 8.

that Johnson was not very eager to reminisce about his early life, and his friends were never too successful in drawing out of him much about his childhood. Boswell's wealth of information was largely concentrated in the later period as described in his own journal entries.

The earliest indication we have that Boswell was thinking of the form his work would take comes in an entry in his journal for October 12, 1780: "I told Erskine I was to write Dr. Johnson's life in Scenes. He approved." [9] But it was not until after Johnson's death in December 1784 that Boswell began seriously to consider the actual writing of the biography. Even then he was slow in starting. Despite the solicitation of his publisher, who was eager to have him provide at once a rival volume to the numerous journalistic lives which were beginning to appear, he refused to be hurried. Nevertheless, he was eager to give the public a taste of what was coming. The obvious answer was to bring out a somewhat revised version of the full journal he had kept during the tour to the Hebrides in 1773, the longest single period during which he had been in close contact with Johnson. With Edmond Malone's invaluable help he did the necessary editing, and the volume appeared in the early fall of 1785. It was an immediate sensation, stirring up intense controversy, largely on ethical grounds concerned with his revelations of Johnson's casual conversation (see pp. 21–22).

Happily, the attacks did not turn Boswell from his main purpose, though they may have induced him to be a little more careful when including material about persons who, like Bishop Percy, were extremely sensitive to publicity. But he was now certain of just what he wanted to do, and in a leisurely fashion he set about the task.

As Geoffrey Scott points out, and as Marshall Waingrow, in his edition of Boswell's correspondence relating to the fashioning of the *Life*, so admirably documents,[10] the process was long and, at times, desultory. In 1786 Boswell was admitted to the English bar, and moved his family to London, where he would be better able to collect material. He interviewed many of Johnson's old friends; he sent a special questionnaire to Edmund Hector, Johnson's old schoolmate, with blanks left for the answers. He gathered all the letters of Johnson he could find. To be sure, Dr. Taylor was loath to let him have those in his possession, and Mrs. Piozzi published separately her large collection. Fanny Burney refused to cooperate, and there were others who were grudging.

[9] *The Private Papers of James Boswell from Malahide Castle*, (hereafter referred to as *Private Papers*), ed. Geoffrey Scott and F. A. Pottle (Mt. Vernon, N.Y.: privately printed, 1928–34), XIV, 132; and *The Correspondence and Other Papers of James Boswell Relating to the Making of the "Life of Johnson,"* (hereafter referred to as "Waingrow"), ed. Marshall Waingrow (New York: McGraw-Hill Book Co., 1969), liii.
[10] Waingrow provides a "Chronology of the Making of the *Life*," pp. li–lxxviii.

Sir John Hawkins, who (much to Boswell's annoyance) had been made the official biographer, had access to Johnson's surviving papers and used them in his own volume. But Boswell got what he could from Frank Barber, Johnson's servant, and most of the Club members, as well as many others, cooperated with Boswell's efforts. Thus he gradually assembled a huge mass of material.

When a first version was finally prepared, it was then rigorously revised, further corrections being made in proof, and some even after sheets were printed. At last *The Life of Samuel Johnson, LL.D* was published on May 16, 1791, exactly twenty-eight years after the first meeting in 1763. It appeared in two large quarto volumes, which sold for two guineas. Widely read and a definite financial success (Boswell received, in all, about £2,500) it stirred up more controversy. In some circles there was the same kind of shocked repulsion that had greeted the Hebridean *Tour*. (For a discussion of this matter, see pp. 22–25.) At the same time it was widely recognized as a remarkable accomplishment in what was seen to be a new literary genre.

Unfortunately, after the appearance of the *Life* Boswell had only four more years to live. His wife had died in 1789, and his own health and spirits were deteriorating. Although he planned other biographies, he accomplished nothing. He saw a second edition of the *Life* through the press and was working on a third when he died on May 19, 1795.

Boswell's Method in Keeping a Journal

Geoffrey Scott, the first editor of the *Private Papers of James Boswell from Malahide Castle,* while unable to study the great mass of documents now available, could in 1929 make some shrewd guesses, which have largely proved correct (see pp. 30–39). More recent scholarly analyses by Frederick A. Pottle and Marshall Waingrow have amplified and filled out the picture (see pp. 45–50, 66–73).

When he came to London in 1762, we now know, the last thing before he went to bed or the first thing in the morning, Boswell was accustomed to write out short memoranda to himself outlining what he intended to do during the day. Along with lists of people to see and things to do, there might be exhortations to himself on how to behave and how to dress. Pottle, in the first printing of *The London Journal,* gives a few characteristic samples.[11] Boswell might begin with some such remark as "Be fine with Macdonald. Think on McQuhae, Countess of Eglinton. Breakfast sunshine, marmalade. . . ." or "Breakfast on fine muffins and [enjoy] good taste of flour. Have hair dressed, and

[11] *Boswell's London Journal, 1762–1763,* ed. F. A. Pottle (New York: McGraw-Hill Book Co., 1950), pp. 199, 211, 223, 249.

if the day is moderate, go to Whitehall Chapel and Lady B's. But if it be cold, stay in comfortable and write journal. . . ." Later in the memoranda there would be names of people to see, with advice to himself as to how to act—"At three, Sheridan's, and be *retenu.* . . ." or "Then go Lady Mirabel's and try siege fairly. Then saunter till five, and then Mrs. Schaw. . . ." or "Then Erskine, then Flexney, then Donaldson; then home and just have tea and bring up fortnight's journal and letters. . . ."

It is clear that every so often Boswell would arrange to stay home and write up his full journal, and no doubt these little memoranda helped to remind him of what had happened. Yet there were obvious discrepancies between what he intended to do and what actually happened. Some days the note might represent a fair outline, but on others it might have been completely modified. For example, on May 16, 1763, the surviving memorandum contains various items about sending out his breeches to be mended, about what to eat for breakfast, about getting money and seeing people, and even what frame of mind to be in, but there is no mention whatsoever of what has been called "the most important single event in Boswell's life"—his meeting with Samuel Johnson.[12]

These memoranda, then, important as they may be to us today, are not reliable as historical data. Although they are valuable in explaining his character, they do not tell accurately what happened. Moreover, they could not have been of too much use to Boswell himself when writing up his journal. Once he recognized this fact, he changed his practice. Normally instead of jotting down notes the night before, or in the early morning, he waited until the end of the day, and then summed up briefly what had actually occurred, sometimes using a kind of shorthand. These notes, as Pottle so clearly shows, were vital to him when he came to write a full account (see pp. 68–70).

Usually Boswell destroyed the original note once the full journal was written up, but happily a few have survived so that his method of expansion can be studied. Scott and Pottle give some good examples and explain the process (pp. 34–6, 68–70). There were times when Boswell never did get around to expanding the notes, and then they remain the only record we have for particular days. Moreover, in later years Boswell was not quite so regular in his expansion of notes to full journal entries. In the later 1770s what survives is often a sequence of very brief notations, some of the most cryptic kind, along with others (the majority) fairly long and quite intelligible, though never as full as those in 1762–63 and in the Hebrides. When Boswell fell behind, instead of going back and trying to catch up, he would write a long

---

[12] *Ibid.,* pp. 259–60.

entry for the day before, or for a few days before, and then cover up the gap with rough notes. After 1776 he preserved very few notes corresponding to the fully written journal, and then only for some obvious reason, as when they might chance to be followed by other notes he never found time to expand.[13] Incidentally, it might be pointed out that Boswell was no meticulous diarist, never missing a day. There are long periods when there are only intermittent entries. Thus sometimes we have fully written journals, sometimes only the condensed, suggestive notes or half-expanded versions, and sometimes nothing at all.

For those who are curious as to the way in which Boswell expanded his original notes, one specific example may be given.[14] On Thursday, June 3, 1784, he was in London, preparing to accompany Johnson to Oxford. The rough note for the day has happily survived [*Private Papers*, VI, 52]:

> Thurs 3 June Restless but not uneasy night fond of M M [his wife, Margaret Montgomerie] hastened to Dr. J. I mentd Sr J Lowthr introd as Peer making mouth water. Bid me think how transient We talked of Langtons inefficiency He repeated from Dryden how the bold impious get—the conscient wait till prize is gone Said Langt taught son to save money Wd send him to Parr if for nothing Talked of Virtue at most never making freindship Good men & bad never enemies —the thing is not disputed Tis Men good different ways. This began by describing the three Langtons wore same livery—same colour of virtue Knotting despicable—Better than nothing—Thats all—As poor employmt for fingers as can be—Tried to learn of Miss Dempster— Garricks distich chace spleen no worse within He tingled wt expect of applause. They did not mind it. Mrs & Miss Beresford—Oxford prayers for dead Ill but wont talk of it. Two lines from Stellas birthday Spectres & pills. If proves cheap people not work—[word undeciphered] pleasures of vanity Bo [defect in MS]

The fully written journal, probably done not too long afterwards, is as follows [*Private Papers*, VI, 55–56]:

> *Thursday 3 June* Had past a restless, but not uneasy night Was full of fondness of M. M. Hastened to Dr. Johnsons as I was to accompany him to Oxford—his first jaunt after his severe illness which confined him so long. I found him calmly philosophical. I told him that Sir James Lowthers introduction into the House of Lords had made my mouth water; and I expatiated warmly on the dignity of a British Peerage. He bid me consider the transient nature of all human honours. I said that would produce indifference. But said he if you allow your mind to be too much heated with desire for a British Peerage, you will

---

[13] I owe this information to Professor Pottle.
[14] From original manuscripts in the Yale Boswell Collection. See *Private Papers*, VI, 52, 55–56. Cf. *Life*, IV, 283–85.

wish to obtain it *per fas aut nefas*. We talked of Langtons inefficiency. He repeated from Dryden some lines which I shall find How the bold impious get. The concientious wait till the prize is gone. He said Langton taught his son himself to save money. He would send him to Parr if he could do it for nothing. I spoke of Langtons Uncle whom I had seen at Rochester. The Dr. said he was a good Man. All the three brothers were good men—Virtuous Men and virtuous the same way they wore the same livery the same colour of virtue. He then made a striking remark—Virtue almost never produces freindship. Good Men & bad are not enemies. They are not embittered by contest. The thing is not disputed between them. Enmity takes place between men who are good different ways.—There is a depth and justness of observation in this remark. Mrs. Dumoulin made tea to us. The Oxford Post Coach stopped at Bolt Court to take us up. Frank had gone off early in the heavy coach. The other two passengers were a widow Mrs. Beresford and Miss Beresford her daughter, the first a very sensible polite Lady the second a genteel pleasing young creature As we passed through Liecester Square I pointed to Sir Joshua Reynolds's house and said—there lives our freind. Ay Sir said the Dr.—there lives a very great Man. I am sorry I do not recollect the particulars of the Drs. conversation But it was so striking that Mrs. Beresford asked me in a whisper If this was the celebrated Dr. Johnson. She had read his *name* at the Coach Office I told her it was. Miss Beresford said to me. Every sentence he speaks is an Essay. I was delighted to observe the immediate effect of his wonderful Genius.

It will be obvious that here some of the original topics were omitted, others developed much further. But this was not the end of the expansion. Some years later, when preparing copy for the *Life* Boswell made other major changes. A comparison with the published book will show what happened.[15] Apparently in this instance Boswell, while using the journal entry, also had occasion to consult the surviving original note, for some things mentioned only in the earlier have found their way back into the published version.

The Question of Accuracy of Reporting

So long as the old theory was accepted—that Boswell was little more than an energetic reporter, writing down what Johnson said—the matter of the verbal accuracy of his versions did not stir up much controversy. Other evidence which was available seemed to confirm his re-

[15] In the printed *Life* Boswell quite reasonably left out the earlier personal details, his envy of Sir James Lowther, and Johnson's remarks about the Langtons. For the biography of Johnson these were not essential. Various items were never used, such as Garrick's distich and the last two items of the note. Johnson's remarks, as later remembered, but not appearing in the journal, are somewhat different in connotation from what is suggested in the note.

ports, and that was that. But now, with the general agreement on Boswell's remarkable creative gifts, there are some questions which need to be discussed. Exactly how did Boswell's memory work? What part of the dramatic scenes in the *Life* represents Boswell's imaginative coloring, and how much is purely factual? Is Boswell's memory more to be relied upon immediately after the event than years later? In other words, is there possibly a gradation of authenticity which depends on when the material was written down in expanded form?

Frederick A. Pottle, who knows more about Boswell than anyone today, believes that Boswell had a very special kind of mind (see pp. 68–70). Once it was given a jog—by a note or in some other way—the whole of an earlier event came back to him in great detail. But he did have to have the jog. Furthermore, it is Pottle's contention that, given the proper reminder, Boswell was just as capable of bringing back little details after ten years as after two weeks. Essentially there is no gradation of accuracy depending on time sequence. Pottle cites various evidence to support his position, including the findings of modern psychologists who have studied people with similar memories.[16] The episode of the ride to Oxford with the Beresford ladies might be taken as a good example. In the original note only their names appear. In the expanded journal entry more is added, but not some of the essential facts. Some six years later, when Boswell was preparing copy for the *Life*, he recalled a great deal more, including the fact that they were Americans, that Mrs. Beresford's husband had been a member of the Continental Congress, and that it was Miss Beresford's "knotting" which had stirred up Johnson's remarks about that occupation. Moreover, he remembered more about Johnson's conversation in the coach. Whenever it has been possible to check these later additions, the details have been shown to be quite accurate. All of this would tend to support Pottle's claim.

On the other hand, some people still have doubts. Is it perhaps significant, they ask, that some of the most colorful and dramatic scenes in the *Life* are those for which Boswell had no complete journal entries? The dinner with Wilkes might be cited as an example.[17] For this he used some of the original notes, which brought back the essential facts, but without any full prior expansion he was freer to fill in the background. Another example is the trip Boswell and Johnson made to Richard Owen Cambridge's on April 18, 1775.[18] Boswell had no full

---

[16] See, for example, F. A. Pottle, "The Power of Memory in Boswell and Scott," in *Essays on the Eighteenth Century Presented to David Nichol Smith* (Oxford: Clarendon Press, 1945), pp. 185–86.

[17] For a discussion of this episode see Sven Molin, "Boswell's Account of the Johnson-Wilkes Meeting," *Studies in English Literature*, III (Summer 1963), 307–22.

[18] *Private Papers*, VI, 40.

journal entry for the day, but his condensed note contains the germ for the delightful scenes later written up for the *Life*. Two samples should be enough to show the method of expansion:

Original note: *Private Papers*	*Life*
. . . Sir Jos good hum—no. Burke no—I look on myself as good hum. . . . (VI, 40)	Johnson. "It is wonderful, Sir, how rare a quality good humour is in life. We meet with very few good humoured men." I mentioned four of our friends, none of whom he would allow to be good humoured. One was *acid*, another was *muddy*, and to the others he had objections which have escaped me. Then, shaking his head and stretching himself at his ease in the coach, and smiling with much complacency, he turned to me and said, "I look upon *myself* as a good humoured fellow." (II, 362)

Obviously Boswell thought it better not to give away the names of any of the friends who had not been thought truly good-humored—a characteristic example of his willingness to suppress some factual evidence if not vital to the story. But what is more important is what it shows us of Boswell's supreme gift of dramatization. The picture of Johnson, complacently stretching himself out in the coach, and remarking on his own good nature, is surely Boswell at his best.

Later in this same note comes the passage:

	Life
Came to Cambrs Gibbons & Burt— Genteel Camb. Johns to Books *Sir Jos* As I to picts but I have advant Can see more than he of books. . . .	No sooner had we made our bow to Mr. Cambridge, in his library, than Johnson ran eagerly to one side of the room, intent on poring over the backs of the books. Sir Joshua observed, (aside,) "He runs to the books, as I do to the pictures: but I have the advantage. I can see much more of the pictures than he can of the books." . . . (II, 364–65)

Some readers may wonder if there are any dependable ways of checking Boswell's reporting of Johnson's conversation. The answer is that there are. Occasionally other people were present who also had the itch to write down what was said by the Doctor, and some of these other reports catch the same "Johnsonian aether." In general when this occurred and the two accounts are compared, there is surprising agreement as to the main ideas expressed, although inevitably there are variations in wording.

Perhaps the best example to be cited is the diary of Dr. Thomas Campbell, an Irish clergyman who came to London in the spring of 1775, and who was with Johnson and Boswell on a number of occasions. Neither he nor Boswell realized that the other was recording some of the conversation, but it is now possible to compare the versions. Campbell's lay hidden many years, until his account turned up, curiously enough, behind an old press in the office of the Supreme Court in New South Wales. A slightly censored version was printed in Sydney in 1854, and a few copies reached England, but a good many people were suspicious of its authenticity because the Johnsonian portions sounded so much like Boswell. Indeed, many wondered if it might not be a hoax, and by the late nineteenth century this suspicion became heightened because by that time the original manuscript had disappeared. It was actually not until 1934 that it turned up again and could be studied by scholars. Now it is possible to compare Campbell's reports with Boswell's for the same occasion. Of course, they did not always record the same remarks, but when they did there is great similarity, and at the same time a few discrepancies.

On April 5, 1775, Campbell dined with Johnson at the Dillys' in the Poultry, as the guest of Boswell. In his diary he set down a number of Johnson's remarks not recorded by Boswell. For instance, there is: "but dinner was then announced & Dilly who paid all attention to him in placing him next the fire said, Doctor perhaps you will be too warm— No Sir says the Doctor I am neither hot nor cold—and yet, said I, Doctor, you are not a lukewarm man.—This I thought pleased him." [19] Note Johnson's "No Sir" which becomes such a characteristic of Boswell's reports.

Only one topic appears in both accounts for this date:

Campbell's Diary	Boswell, *Private Papers*
Talking of Addison's timidity keeping him down so that he never spoke in the house of commons was he said much more blameworthy than if he had attempted & failed; as a man is more praise worthy who fights & is beaten than he who runs away. (p. 75)	We talked of speaking in Publick. Mr. Johnson said that one of the first wits of this Country, Isaac Hawkins Brown, got into Parliament and never opened his mouth. Mr. Johnson said that it was more disgraceful not to try to speak than to try and fail, as it was more disgraceful not to fight than to fight and be beat. . . . (X, 188)

[19] Dr. Campbell's Diary of a Visit to England in 1775, ed. J. L. Clifford (Cambridge: Cambridge Univ. Press, 1947), p. 73.

The fact that Campbell cites Addison as the timid figure and Boswell Isaac Hawkins Browne need not indicate an error for either commentator. Johnson may well have cited both men.

On April 8 Campbell was again with Johnson and Boswell at the Thrales' in Southwark. Again they recorded some of the same remarks.

Campbell's Diary	Boswell, *Private Papers*
Boswell lamented there was no good map of Scotland.—There never can be a good of Scotland, says the Doctor sententiously. This excited Boswell to ask wherefore. Why Sir to measure land a man must go over it; but who cd. think of going over Scotland? (p. 76)	I told him that Mr. Orme said many parts of the East Indies were better mapped than the highlands of Scotland. Said Mr. Johnson: "That a country may be mapped, it must be travelled over." "Nay," said I, "can't you say it is not *worth* mapping?" (X, 213)
Mrs. Thrale then took him by repeating a repartee of Murphy—(The setting Barry up in competition with Garric is what irritates the English Criticks) & Murphy standing up for Barry, Johnson said that he was fit for nothing but to stand at an auction room door with his pole &c—Murphy said that Garrick wd. do the business as well & pick the people's pockets at the same time.—Johnson admitted the fact but said Murphy spoke nonsense for that peoples pockets were not picked at the door, but in the room &c &c—Then say'd I he was worse than the pick pocket, forasmuch as he was Pandar to them—this went off with a laugh—*vive la Bagatelle.* (p. 77)	Mrs. Thrale told us that Mr. Johnson had said that Barry was just fit to stand at the door of an Auction-room with a long pole: "Pray, Gentlemen, walk in." She said Murphy said Garrick was fit for that, and would pick your pocket after you came out. Mr. Johnson said there was no wit there. "You may say of any man that he will pick a pocket. Besides, the man at the door does not pick pockets. That is to be done within, by the Auctioneer." (p. 199)

One could go on and on citing examples. The main point in each instance is the same, only the peripheral matter is different, and this depends on the narrator's own particular interests and natural desire to put himself into the picture. Thus the comparison does reassure us of the general reliability of Boswell's reporting of Johnson's ideas, and at the same time gives further evidence of his dramatic skill.

Gathering Material for the Life

In Marshall Waingrow's remarkable edition of that part of Boswell's correspondence which has to do with the fashioning of the *Life,* one can see exactly how he proceeded. That Boswell had a keen interest in securing accurate facts has never been doubted. As he bragged in the "Advertizement" to the first edition of the *Life,* he was quite ready to "run half over London" to verify a date. And it is now evident that he used admirable skepticism in regard to casual anecdotes which he secured from others. While what he received may have been firsthand to those who gave him the stories, it was secondhand material for Boswell, and he treated it with suspicion. Since he cast his net wide for information, it was inevitable that he would draw in much that was dubious. Thus on one letter he received from Oxford we find Boswell's pungent note: "Nonsense about Dr. Johnson." Refusing to accept doubtful contributions, even from Johnson's intimate friends—George Steevens among them —Boswell remained a strict censor. When he grew suspicious of some of Anna Seward's stories, he commented to Mrs. Cobb of Lichfield, who had been inaccurately cited as the source of one of them, "As I find my authority quite erroneous in one remarkable particular, I cannot trust to it for any part." [20]

Only occasionally was he led astray. In one instance Edmund Hector, who was his chief authority about Johnson's early years, forced him to reject a hint he had had about Johnson's having been for a short time just before his marriage a tutor in the household of Thomas Whitby of Great Haywood.[21] Hector unfortunately on this point was wrong, for the story, we now know, was correct.

Where his chief rivals, Mrs. Piozzi or Sir John Hawkins, were concerned, Boswell tended to be a little less careful. He was quite willing to accept material from Giuseppe Baretti, who was not particularly noted for scrupulous accuracy, when it supported his own instinctive suspicions of Mrs. Piozzi's veracity. One example may show the kind of problems which face modern experts when attempting to evaluate the evidence.

In her *Anecdotes of Dr. Johnson* Mrs. Piozzi, after giving examples of Johnson's occasional rudeness, added:

> He was no gentler with myself, or those for whom I had the greatest regard. When I one day lamented the loss of a first cousin killed in America—"Prithee, my dear (said he), have done with canting: how would the world be worse for it, I may ask, if all your relations were at once spitted like larks, and roasted for Presto's supper?" [22]

Boswell in the *Life,* after commenting on Mrs. Piozzi's exaggeration

[20] Waingrow, p. 288. See also p. 173, note 19.
[21] *Ibid.,* p. 172.
[22] *Johnsonian Miscellanies,* ed. G. B. Hill (Oxford: Clarendon Press, 1897), I, 189.

and distortion and her tendency to paint Johnson as deficient in
tenderness and ordinary civility, continued:

> I allow that he made her an angry speech; but let the circumstances
> fairly appear, as told by Mr. Baretti, who was present:
> 'Mrs. Thrale, while supping very heartily upon larks, laid down
> her knife and fork, and abruptly exclaimed, "O, my dear Mr. John-
> son, do you know what has happened? The last letters from abroad
> have brought us an account that our poor cousin's head was taken
> off by a cannon-ball." Johnson, who was shocked both at the fact,
> and her light unfeeling manner of mentioning it, replied, "Madam,
> it would give *you* very little concern if all your relations were spitted
> like those larks, and drest for Presto's supper." ' [23]

This is the version which most readers since that time have accepted
as correct.

It happens that years later Mrs. Piozzi carefully annotated at least
two different editions of the *Life*. Opposite this story she wrote in
the margins such comments as: "Boswell appealing to Baretti for a
Testimony of the *Truth* is comical enough"—"I never address'd him
so familiarly *in my Life*. I never did eat any Supper:—& there were
no Larks to eat"—"Nor was ever a *hot dish* seen on the Table after
Dinner at Streatham Park." [24] She did not deny the basic fact of her
remark and Johnson's reply, but all the little details which rendered
Baretti's account so damning would seem to have been figments of his
own imagination. Apparently Boswell never thought it worthwhile
to ask anyone about the regular eating habits of the Streatham household.

At times the lady confirmed what her rival was saying, as when
Boswell tells of Johnson's later fond recounting of his wife's approba-
tion of the *Rambler,* and she noted in the margin: "he told me the
same thing in the same Words."

We now know that Boswell was subjected to pressures of various
sorts to keep him from including full details about certain of Johnson's
transactions. Roger Lonsdale has recently shown, in an important
study based on manuscript material in the Osborn and Hyde collections, how Dr. Burney censored Johnson's letters to him before
giving copies to Boswell.[25] In this instance, then, tampering with the
letters as they appear in the *Life* cannot be charged to Boswell. He
printed what he was given.

[23] *Life,* IV, 347.
[24] *Life,* ed. Edward G. Fletcher (London: Limited Editions Club, 1938), III, 401. Included are Mrs. Piozzi's annotations in the 8th edition (London: 1816), now at Har-
vard, and in the 5th edition (London: 1807), now in the Hyde collection.
[25] "Dr. Burney and the Integrity of Boswell's Quotations," *Papers of the Bib-
liographical Society of America,* LIII (4th Quarter 1959), 327–31.

In 1784 the ailing Johnson had drafted a dedication to the King for Burney to use in his *Account . . . of the Commemoration of Handel*. As Lonsdale puts it,

> Three years later, in 1787, Burney's anxiety to be commemorated in Boswell's great work as an intimate friend and regular correspondent of Johnson was matched by an equal anxiety to conceal Johnson's authorship of the noble Dedication. Burney solved the problem by providing Boswell, in the last three pages of his narrative, with partial transcripts of these letters, omitting all but the most innocent mention of the *Commemoration* and Johnson's part in the work.

It is clear, too, that Burney omitted at least four letters from the list he gave Boswell, perhaps because one of them clearly referred to help Johnson was then giving Burney with the second volume of his *History of Music*.

Nor was Dr. Burney the only one of Johnson's close friends who put pressure on Boswell to suppress evidence. Sir Joshua Reynolds was just as eager that there be no mention in the *Life* of various assistance he had had from Johnson.

The wealth of evidence now available does make clear one important point: Boswell did not use all the material he collected, even when it was particularly relevant. Occasionally he did some slight censoring. Thus he omitted an amusing story told him by Hector of a night in Birmingham when Johnson as a young man may have been drunk.[26] Apparently Boswell thought the story might give a wrong impression about his hero. Another example has to do with the relations of Johnson and his wife Tetty. Perhaps because Boswell did not wish to destroy the accepted view of Johnson's sentimental devotion to the memory of his wife, so conclusively shown in the great man's later prayers, Boswell decided to omit any mention of Johnson's decision, a year after Tetty's death, to seek a second wife.[27] And there was a fascinating interview which Boswell had with Mrs. Desmoulins which was largely concerned with the question of Johnson's sexual capacity.[28] Obviously, this was much too controversial to be included. Other small details which Boswell felt were either irrelevant or not characteristic of the great moralist he was describing were silently omitted.

This does not mean that Boswell was consciously distorting character or falsifying evidence. Like all great biographers, he was presenting the essential truth as he saw it. Critics at various times have pointed

[26] Waingrow, pp. 91 and xliii. Also J. L. Clifford, *Young Sam Johnson* (New York: McGraw-Hill, 1955), pp. 142–43, 310–16, etc.

[27] Donald and Mary Hyde, "Dr. Johnson's Second Wife," *New Light on Dr. Johnson*, ed. F. W. Hilles (New Haven, Conn.: Yale Univ. Press, 1959), pp. 133–51.

[28] Yale Boswell Papers, 20 April 1783; *Young Sam Johnson*, pp. 313–15.

out that it was not humanly possible for Boswell to show us the
whole Johnson (see pp. 79–89, 97–111, 113–14). There were other sides
of the man which he had little opportunity to see. Most of Boswell's
contacts with his subject, at the Club or dining in other people's houses,
were largely masculine in character. He rarely saw him with children,
or in the easy domesticity of a family. Thus Mrs. Thrale and Fanny
Burney described other aspects of the great man, and rounded out
the picture. One other side of Johnson's character, which Boswell saw
fit to play down, was his capacity for simple fun. The stories of
Johnson boisterously climbing trees, or rolling down hillsides, or
running races barefoot on the lawn, come from other observers.

The *Life*, then, gives us Boswell's Johnson, and its strength lies
in this very point. In order to show the reader what Johnson meant
to him, Boswell necessarily had to make some choice of material.
The great conversations, so carefully set down in his journals, were
Boswell's most important contribution, and he was wise enough to
recognize this fact. These no one else could ever equal. Thus he
deliberately made them the core of his work.

The Manuscript of the Life

The persistent question of how Boswell drew all this diverse material
together can now be answered with certainty. In 1929 Geoffrey Scott
was forced to judge from very scant evidence. Having only a few
scattered pages of what he assumed to be the manuscript of the *Life*,
and believing that the rest had all crumbled away or been destroyed,
he carefully considered all possible hypotheses, and with brilliant
intuition finally arrived at a theory which has since been shown to
be substantially correct.

Shortly after Scott's untimely death and the assumption of editorship of the Malahide Castle Papers by Frederick A. Pottle, more
evidence became available—some 120 quarto leaves of the manuscript
of the *Life* (in the possession of Mr. Arthur Houghton). With such
a long sequence, Professor Pottle was able in 1931 to substantiate
Scott's conjectures and to give the best analysis so far in print of
Boswell's method.[29] Yet even then Pottle was forced to theorize from
only a small percentage of the whole work, and much of the interpretation had to be guesswork. So it remained until the 1940s. Then in the
stable loft at Malahide Castle there turned up more than 900 quarto
leaves of the actual manuscript sent to the printer, as well as most

[20] F. A. and Marion S. Pottle, *Catalogue of the Private Papers* . . . (London: Oxford Univ. Press, 1931), item 303; and various other essays by Pottle.

of what Boswell called "Papers Apart." With only a few gaps, it is now possible to reassemble all of the actual copy which the printer used in setting up the two volumes of the 1791 edition. Someday it will be published under the editorship of Herman W. Liebert and Marshall Waingrow. Meanwhile, a short description may be of some value.[30]

Boswell carefully wrote his first rough draft on one side of quarto leaves of uniform size. Letters, passages from printed sources, other miscellaneous notes and documents, however, were not copied into the main narrative but were kept in separate piles as "Papers Apart." These he wove into the main narrative by the use of an elaborate system of reference symbols and notes to the printer. "Go to Paper △△," he would write, or "Take in his of 14 March," or "Excerpt my letter 19 April." Apparently Boswell kept the printer supplied with stacks of these "Papers Apart," which had to be carefully dovetailed into the manuscript at the proper point. Sometimes there are harried notes from the printer to Boswell indicating that the press was at a standstill and asking for more copy. Included also in the "Papers Apart" are large sections of his journals which Boswell intended to print almost verbatim. In the Yale collection these have now been returned to their proper place in the journals, but it is possible to fit them easily into the intricate plan of the work. It might be added that of these "Papers Apart"—made up of journal pages, tiny scraps and notes, separate documents, and printed works—almost everything has survived except the books from which some of the printed selections were taken and the originals of Johnson's letters to Boswell.

The complete draft, not only of the main narrative but also of the material taken from the "Papers Apart" was later subjected to the most stringent revision. Not merely stylistic, this meticulous correction encompassed in a large part the whole matter of arrangement of material and the presentation of it. It was a major creative effort, almost an entire rewriting of the manuscript. Boswell at the start had written on one side of the quarto leaves, with the verso reserved for later additions and corrections (that is, the verso of the preceding leaf, when the leaves were laid open, served as a catchall for revisions and additions to the following page). By the time the final version was complete, many of these versos were completely filled and very few were blank. If anyone still has the mistaken notion that Boswell was

[30] What follows is largely taken from a paper read by the present editor at the M. L. A. meeting in December 1950. My thanks are due for generous help given then by Professor and Mrs. Pottle, Mr. Herman W. Liebert and Professor Marshall Waingrow.

artistically lazy, that he merely joined together a mass of fragments and called it a biography, let him examine the minute revisions and major rearrangements made during the revision of his first draft.

In the past some scholars have wondered how much credit should be given to Edmond Malone for whatever merits in form and organization the work may have. Malone's intimate connection with the preparation for the press of the *Journal of a Tour to the Hebrides* is well known, and it is sometimes assumed that he did the same kind of editorial work on the *Life*. Actually his handwriting appears only in a few places in the manuscript of the *Life*. To be sure, this does not mean that Malone's good sense was not available to Boswell when he was correcting the first draft. We know from Boswell's journals that as long as Malone was in London, particularly during the last months of 1789 and early 1790, Boswell did most of the revising at Malone's house. Again and again he was to write in his journal that he dined with Malone and revised the *Life*.[31] Thus we can picture Boswell, carefully rewriting his first version, asking Malone's advice on this problem or that, depending on his friend's critical judgment for excisions or additions. But in the last analysis it was Boswell who made the changes and who prepared the final complicated draft for the printer.

Until the manuscript is finally scrupulously edited, only a cursory analysis is possible. A modern scholar attempting to examine the material experiences at first complete bewilderment. The intricacy of deletions and changes makes the narrative line difficult to follow, particularly if one is attempting to sort out the levels of revision. The mere physical form of the papers is often amusing. In one instance a series of pages was fastened together to make an accordion-like extension almost a yard long. How any printer could have worked from such copy remains a mystery. But with Boswell at his elbow, and with no union restrictions, the printer did get the copy set, and a masterpiece resulted.

Merely to attempt to follow the devious trail as the sections come together is in itself an adventure. It is entertaining, to begin with, to watch the biographer communing with himself as he goes along. Here in the margin is a note to himself asking if he has already related an anecdote in another place, or perhaps reminding himself to check quotations or to fill in blanks later with names and dates. Once after quoting Johnson on the "ridiculous" character of a friend, leaving the man unidentified, Boswell wrote in the margin a question as to whether or not it would be better to tell it specifically of Langton. In this instance kindness finally won and he did not. Boswell could

[31] See "Chronology" in Waingrow, pp. li–lxxviii.

not refrain from amusing off-the-record comments, and often seemed
almost to be looking over his own shoulder as he wrote. Opposite an
unusually flattering notice of Capell Lofft, Boswell commented:
"There *must* be *some* touches for popularity." Later, of course, once
decisions had been made, all these notes to himself were crossed out.

The first draft of the *Life* appears to have been written with the
same fluency as the journals, with little hesitation or uncertainty,
with not much searching for what to say or recasting of sentence
structure. To be sure, the speed was partly the result of his putting
off decisions as to choice of words and phrases, for he frequently introduced alternatives with the idea of making a choice when he came
back to revise. He sometimes enclosed phrases within virgules, indicating that he was uncertain which to choose. Later either the
words or the virgules would be deleted. When he came to the major
revision Boswell continued to experiment with phraseology. The
order of words might be reversed, modifiers changed, and whole
sentences rewritten. Boswell's ear was sensitive to modulations of
sound within a sentence. "Loved and caressed by everybody" in the
earlier draft becomes "caressed and loved by all about him."

In making changes of wording Boswell was always striving for more
color, as well as more precision. "Remarkably lively and gay and
very happy" consequently becomes "a gay and folicksom fellow."
Sometimes it appears that he reaches his final version almost by
trial and error. For example, Boswell tells of a dispute he had with
Mrs. Thrale as to "whether Milton or Shakespeare had given the
best portrait of a man." In revision he reverses the two names, making
it "Shakespeare or Milton"; "given" becomes "drawn," and "best
portrait" first becomes "noblest description" and then finally "most
admirable picture." For the most part, the changes do improve the
style, though we may lament an occasional change such as the altera-
tion of "was very angry" to read "expressed his disapprobation of."

Boswell's struggles with words usually come in his own descriptive
passages, rather than in the transcripts of Johnson's conversation,
though there is some shaping of the great man's actual remarks.[32]
When Boswell is remembering an episode long afterward there is
occasionally some indecision.

For those who have the impression that Boswell put into the *Life*
everything he knew about Johnson, a study of the manuscript will
be a sobering experience. Boswell had available many good stories
of his hero which, with stern discipline, he rigorously refused to use.
To be sure, not everything crossed out was omitted. A few passages

---

[32] For one example, compare *Life*, I, 284 with Waingrow, p. 24. "Damn Maty—
little dirty-faced dog" in Adams' account is changed to read "*He,* (said Johnson)
the little black dog!"

were merely transposed to other sections of the biography. Boswell
did some reshaping of structure of this sort. Nevertheless, the amount
of material actually taken out of the text in his final revision is astonishing.

What kind of things, it may be asked, did he cut out? They fall
into several clearly defined categories. In the first place, there were
details about his own personal affairs which Malone or his own good
sense dictated that he omit. Not sensational, they are concerned
with matters not essential to a true understanding of Johnson's
character. It could hardly have been of much use for posterity to know
that one evening in Miss Williams' room Boswell had been entertained
with oysters and porter. Likewise in revision he took out various
arguments about conflicting evidence, as when he had balanced the
opposing opinions of Taylor and Hector concerning Johnson's early
intellectual development. Also he cut out a long, acrimonious attack
on his rival Mrs. Piozzi, which Courtenay had convinced him he
should lighten for his own credit. It is amusing to discover that
Boswell originally had planned in the first edition of the *Life* to
quote from his own ribald "Ode by Dr. Samuel Johnson to Mrs.
Thrale upon Their Supposed Approaching Nuptials."

Perhaps the most interesting passages for us are those which were
omitted obviously for prudential reasons because they involved other
people who were still living. For example, a very amusing anecdote
about Johnson's reaction to a rather rude remark of a hard-drinking
Captain Brodie, who had married Molly Aston, was taken out possibly
to avoid hurting the Captain's feelings. References to close friends
like Bennet Langton were either omitted as being perhaps too cutting,
or were told anonymously.[33] And there were other anecdotes which
Boswell, after consideration, must have decided were dubious, or which
conflicted with better authenticated facts.

It might be added that the process of pruning did not end with
the preparation of the manuscript for the printer. There were further
excisions in proof, and in the last stages, even after the sheets were
printed. One example is an anecdote which Boswell had from Johnson concerning an interchange with Tetty about conjugal infidelity,
and her remark that she "did not care how many women he went
to if he *loved* her alone." Boswell originally included this in the
Life, but was finally persuaded by Windham and others to take it
out, even though it meant cancelling a printed page. As he wrote to
Malone to tell him of the decision, he had bowed to the advice of
others, but added, "It is however mighty good stuff." [34]

[33] See L. F. Powell's "A Table of Anonymous Persons" where many are identified,
in *Life,* VI (2nd ed.), 431–75.
[34] Waingrow, p. 384.

There is the inevitable question: Would the *Life* have been an even greater book if Boswell had not done all this revising and censoring? The answer must be a qualified one, depending upon which point of view one takes. Except for certain omissions made for prudential reasons, most of the reshaping did improve the *Life* as a work of art. And this testifies to Boswell's literary concerns and judgment. As students of Boswell and Johnson we may welcome all the new evidence contained in the deleted passages, but as students of the art of biography we must applaud Boswell's skill in producing a smooth, engrossing narrative.

Contemporary Reception of Boswell's Johnsonian Works

Because the *Tour to the Hebrides* and the *Life of Johnson* have for so long been recognized as masterpieces in their genre, we sometimes forget that when they first appeared they were subjected to the same kind of abuse that is leveled against some candid biographers today. Boswell's frank reporting of actual conversations (including Johnson's occasional adverse opinions of others) without securing the permission of those involved, aroused instant revulsion.[35] This was simply not done in polite society. Thus the works were subjected to the same kind of vigorous attack as have been Lord Moran's revelations concerning Winston Churchill's illnesses, or William Manchester's gossip about the Kennedys.

The *Tour* when it appeared in 1785 was constantly under attack in the newspapers: on October 1 the *Morning Post* commented "Had Dr. Johnson been blessed with the gift of *second-sight*, how it would have tortured him to have known the base advantages which have been taken of his celebrity to make money." On through October and November the attacks continued. Horace Walpole called the *Tour* a "most absurd enormous book. . . . the story of a mountebank and his zany," and Michael Lort reported that Edmund Burke had fallen hard upon Boswell for the many absurdities in the book. John Wilkes is said to have told Boswell that "he had wounded Johnson with his pocket pistol & was about to despatch him with his blunderbuss when it should be let off." [36] The popular satirist Peter Pindar found Boswell's firsthand reporting a perfect target:

[35] For a discussion of what was involved see J. L. Clifford, "How Much Should a Biographer Tell? Some Eighteenth-Century Views," in *Essays in Eighteenth-Century Biography*, ed. Philip B. Daghlian (Bloomington: Indiana Univ. Press, 1968), pp. 67–95.
[36] Walpole, *Correspondence*, ed. Toynbee, XIII, 337 (6 October 1785); letter of Michael Lort to H. L. Piozzi, (31 December 1785) [John Rylands Library MS. 544–5].

I see thee stuffing, with a hand uncouth
An old dry'd whiting in thy Johnson's mouth;
And lo! I see, with all his might and main,
Thy Johnson spit the whiting out again.
Rare anecdotes! 'tis anecdotes like these,
That bring thee glory, and the million please!
On these shall future times delighted stare,
Thou charming haberdasher of small ware.[37]

In fashionable and conservative circles what shocked many readers was Boswell's willingness to report exactly what people said in private conversation. It is reported that Lord Monboddo, when asked what he thought of Boswell, replied: "Before I read his Book I thought he was a Gentleman who had the misfortune to be mad; I now think he is a mad man who has the misfortune not to be a Gentleman." [38] Mrs. Montagu, "Queen of the Bluestockings," in a letter to Mrs. Piozzi, castigated Boswell for his disclosures. "Would any man who wish'd his friend to have the respect of posterity exhibit all his little caprices, his unhappy infirmities, his singularities?" [39] James Beattie summed up this attitude in a letter to Sir William Forbes: "Johnson's faults were balanced by many and great virtues; and when that is the case, the virtues only should be remembered, and the faults entirely forgotten." [40] Biography according to this position should embalm, not re-create.

Of course, there was some praise for the *Tour* in the journals. Everyone realized that Boswell had shown phenomenal skill in delineating his subject. No one doubted the factual truth of his account, only the ethical justification for such revelations. And on this point Boswell was almost universally denounced.

On the appearance of the *Life,* almost six years later, there was the same kind of shock over the frankness of the personal revelations it contained. We now know that Boswell tried hard not to hurt the feelings of Johnson's close friends, and had been willing to do some significant censoring, but none of this was readily apparent to the ordinary reader. On the surface it appeared that he was indiscreetly telling everything. And many thought that such lack of taste not only hurt others, but was injurious to the reputation of Johnson himself. To pass on to the public remarks which a man may have made in

[37] *A Poetical and Congratulatory Epistle to James Boswell* (London: 1786).
[38] B. R. McElderry, Jr., *Notes & Queries,* July 1962, p. 268.
[39] See Clifford, "How Much Should a Biographer Tell?" p. 87.
[40] Sir William Forbes, *Life of James Beattie* (London: 1806), II, 184. Boswell may be said to defend himself against such criticism in the *Life,* I, 30–34. See also *Life,* V, 238.

casual conversation, and to describe the unimportant events of day-to-day existence, was thought to be too great a violation of privacy. Instead of being instructive and edifying, the new biography appeared to be a mere gratification of impertinent curiosity.[41]

Anna Laetitia Barbauld commented: "It is like going to Ranelagh; you meet all your acquaintance: but it is a base and a mean thing to bring thus every idle word into judgement—the judgement of the public." [42] As Samuel Whyte later put it, "A great character, in worthy situations, is an object of virtuous contemplation; but that minuteness of anecdote, that ostentatious display of trifles, which we sometimes meet with, is a vicious indulgence of inquisitive impertinence; a flagrant breach of private confidence, and an infringement of the rules of good breeding." [43] Boswell, Bishop Percy insisted, "by publishing private and unguarded conversation of unsuspecting company into which he was accidentally admitted," had violated one of the "first and most sacred laws of society." [44] Charles Blagden called the *Life* a new kind of libel, by which it was possible to abuse anyone by attributing the abuse to someone who was dead.[45]

There is even some indication that in later years Boswell himself was not graciously received in certain circles, for fear that he would write down what was said and perhaps publish it. Percy once remarked that Boswell had been "studiously excluded" from decent company, and the wife of Archibald Allison, who was upset by what she thought Boswell's "gross gossipation," commented "how well he deserves what he daily meets with that of people shutting their doors against him as they would against any other wild Beast." [46] But such a reaction represented the attitude of only certain levels of society.

At the same time, many general readers were quite ready to accept the new approach with enthusiasm. Ralph Griffiths in the *Monthly Review* discussed the whole question of how much evidence a biographer should include. His reply to those who had been objecting to the new overall coverage was unequivocal.

> On the other hand, an approver will contend, that where the biographer has for his subject the life and sentiments of so eminent an instructor of mankind as Samuel Johnson, and so immense a store-house of mental treasure to open and disclose to the eager curiosity of rational and

---

[41] See, for example, Vicesimus Knox, *Winter Evenings*, 2nd ed. (London: 1790), I, 105, etc.

[42] Anna Laetitia Barbauld, *Works*, ed. Lucy Aiken (London: 1825), II, 157–58.

[43] Samuel Whyte, *Miscellanea Nova* (Dublin: 1800), pp. vi–vii.

[44] Robert Anderson, *Life of Samuel Johnson*, 3rd ed. (Edinburgh: 1815), p. 6.

[45] See *Life*, IV, 30, note 2.

[46] See McElderry (note 38 above); also Waingrow, pp. 237, 435ff.

laudable inquiry, there can be no just exception taken against the number and variety of the objects exhibited. He will ask, 'What conversation could have passed, where so great a genius presided, at which every man of learning and taste would not wish to have been present, or, at least, to have it faithfully reported to him?' To the reporter, would he not say 'Give us *all;* suppress nothing; lest, in rejecting that which, in your estimation, may seem to be of inferior value, you unwarily throw away gold with the dross.'

Griffiths further insisted that he was among the readers of Boswell

who do not think that he has set before us too plenteous an entertainment; nor have we found, that, often as we have sat down to his mental feast, we have ever risen from it with a cloyed appetite.[47]

And others felt the same way. Boswell's friend Wilkes told him it was "a wonderful book," and James Beattie thought it a "great work." [48] A reviewer in the *Gentleman's Magazine* summed up the achievement this way: "A literary portrait is here delineated; which all who knew the original will allow to be the MAN HIMSELF," and another writer in the same periodical added: "no book that has appeared in this age deserves better the popularity which it has already obtained, and which will undoubtedly increase." [49]

Elsewhere I have shown how little significant criticism there was concerning the art and ethical principles of biography before the late eighteenth century.[50] Life-writing had simply not been accepted as a major literary genre, and as a minor adjunct of history it was not thought to be a proper subject for extended discussion. It is a startling fact that in the thousands of periodical essays during the first half of the eighteenth century, not one provided any important critical evaluations of the technical problems of a biographer. Johnson's *Rambler* essay No. 60 broke new ground in 1750, but even this did not go deeply into the ethical difficulties connected with the re-creation of the life of a person recently dead. It was not until after the extended arguments stirred up by Boswell's thorough and revealing coverage of Johnson's private life that biography gradually took its place as one of the important kinds of writing worthy of searching critical examination.

With the argument over Boswell's *Life of Johnson* the whole issue

[47] *Monthly Review,* n.s. VII (January 1792), 3–4; B. C. Nangle, *Monthly Review: Second Series* (Oxford: Clarendon Press, 1955), p. 92.

[48] *Letters of James Boswell.* ed. C. B. Tinker (Oxford: Clarendon Press, 1924), II, 437; Waingrow, p. 482 (letter from Beattie to Boswell, May 3, 1792).

[49] *Gentleman's Magazine,* LXI (May and June 1791), pp. 466, 499–500.

[50] See Clifford, "How Much Should a Biographer Tell?"

as to how much a biographer should tell was finally brought into the open. Here was something on which critics might continue to disagree. Even though the next century kept insisting on reticence and good taste, the possibilities of three-dimensional re-creation of character, both psychological and factual, were now apparent.

Later Criticism

To evaluate the effect on later generations of Boswell's formula for biography is not easy. In one respect the intimacy of his revelations suited the romantic stress on the individual and his inner motives, but ran counter to the gradually maturing reticence and emphasis on decorum of the nineteenth century. Thus his full picture of Johnson as a person could be applauded, while his gossip about others and his reporting of casual conversation were deplored. The result was qualified acceptance of the aims of the new biography, but a refusal to follow Boswell in detail. Characteristic is Lockhart and his life of Scott. While influenced greatly by the form of the *Life of Johnson*, Lockhart steadfastly refused to Boswellize his subject.

In a recent book Joseph Reed, Jr. has examined the various reasons for the refusal of later biographers to follow explicitly Boswell's example, and at the same time has commented on the tremendous expansion of life-writing as a popular genre in the early nineteenth century.[51] There is no need to repeat all the evidence here. Biography had undoubtedly come into its own, though external circumstances kept it from developing in the direction begun by Boswell. The Victorian "Life and Letters" tradition, while owing much to his technique, also went back to the commemorative tradition of earlier centuries.

As suggested earlier, the *Life of Johnson* thus became valued more as a mine of information (Croker's edition for example) and as a source of amusement for readers than as a genuine work of art. Only recently has there been any major study of Boswell's techniques. Today we can argue at length over his possible limitations (see part two), over his method of organizing his material, his excellence as a dramatist, his use of aesthetic distance, and his general artistic accomplishment (see pp. 27–30, 45–78, 90–96, 112–15). The *Life of Johnson* has finally been

---

[51] Joseph W. Reed, Jr., *English Biography in the Early Nineteenth Century, 1801–1838* (New Haven: Yale Univ. Press, 1966). See also Francis R. Hart, "Boswell and the Romantics: a Chapter in the History of Biographical Theory," *ELH*, XXVII (March 1960), pp. 44–65.

accepted as a masterpiece of literature. What follows attempts to provide a sampling of the various modern approaches.

I should like to acknowledge the help given me by the various scholars represented in the volume, and also by Professor Maynard Mack, the General Editor of the Series, and by Professors Maurice Quinlan, Marshall Waingrow, Frederick A. Pottle, and my wife.

PART ONE

Boswell's Materials and Techniques

The Making of the Life of Johnson as Shown in Boswell's First Notes

by Geoffrey Scott

The First Records

I

The *djinn* in the Arabian tale, once liberated from the confining vessel, forms and expands his astonishing shape, until at length the whole chamber is filled, and men look on at his vast and unpredicted motions. So, since 1791, when the world first opened Boswell's book, the figure of Johnson is enlarged, has filtered through our air and yet is still palpable, has achieved not mere pale immortality but an increase of demonic life. Time, which has lent him the force of a symbol, has not lessened his actuality as a man.

When we examine the book from which this urgent figure is risen, we find a close mosaic of small separate facts and sayings. No biography was ever so free from generality; there is no attempt to explain the secret, to forestall the shape that will form itself on the air; scarcely any propounding and summing; all is particular. Boswell weighs out each tested fragment; and the speck of radium inhering in each generates the energy by which the great total, Johnson, strides on among the living.

That there is sensitive art in the weighing out, is plainly evident. But Boswell's conscious effort seems to be fixed far less upon art than upon authenticity. In his letters and diaries we overhear the

---

"*The Making of the* Life *of* Johnson *as Shown in Boswell's First Notes*" by *Geoffrey Scott. From* The Making of the Life of Johnson, *vol. 6 of* Private Papers of James Boswell from Malahide Castle in the Collection of Lt. Colonel Ralph Heyward Isham, *ed. Geoffrey Scott and F. A. Pottle (Mt. Vernon, N.Y.: privately printed, 1929). Reprinted with permission of Yale University and the McGraw-Hill Book Company.*

groans of authorship; but we are witnessing the contrition of an idler or the perplexities of a scholar, never the doubts, still less the despairs, of an artist. Boswell shrank at times from the sheer material magnitude of his task; he worried over his financial profits; above all he tortured his friends and himself in the effort to gather his harvest of particulars, and he will run half over the city to verify a date. But once at work, never does he question how to give "effect" to this or that element of humour or poignancy, or whether he can convincingly balance the light and shade in Johnson's character. To collect enough facts, and (since nothing less than all can be enough) to collect more, and to be satisfied of their authenticity: these are his anxieties. Of his power to give life to the vast pile he never hints one doubt.

Yet he knew the best he could do must still fall short of that platonic standard, the *idea* of Johnson, laid up, in those who knew him, incommunicably, behind all words. To entertain and move his hearer with a "rendering" of Johnson, did not satisfy his high sincerity. To create a living figure might still fall short of truth. Boswell has an image which describes his aim: a "life" should be like a flawless print struck off from the engraved plate which is bitten in our memory. Truth to that archetype must be not only line for line, but tone for tone. Biography should be nothing less than this *duplication* of an image in the mind; not a selection or a monument or a thesis. An aim beyond human reach. The knowledge that his arrow pointed to that impossible mark, was Boswell's source of confidence. Other biographers might *forestall* his book; that they could rival it, he never, in his most sunken moments, conceived. Those others did not even know that biography is impossible.

Thus Boswell's fine assurance that he could do the work better than another man is closely coupled with the modesty of that knowledge of a fixed and remote aim. On October 19, 1775 he drove into Edinburgh in a chaise with his Uncle, Commissioner Cochrane. In his diary he wrote, *"The great lines of characters may be put down. But I doubt much if it be possible to preserve in words the peculiar features of mind which distinguish individuals as certainly as the features of different countenances. The art of portrait painting fixes the last; and musical sounds with all their nice gradations can also be fixed. Perhaps language may be improved to such a degree as to picture the varieties of mind as minutely. In the meantime we must be content to enjoy the recollection of characters in our own breasts . . . I cannot pourtray Commissioner Cochrane as he exists in my mind."*

From such a private reflection as this, one can see how conscious was Boswell's art of biography; how habitually aware he was that the ideal biography had never been written, or even attempted. "The great lines," the method of Walton, might give you an ennobled statue; it needed the minute particulars of Plutarch to make a man. But how few and insufficient are Plutarch's particulars, how inauthentic or remote beyond possibility of verification; how casual a harvest; how far short, in effect, of "Commissioner Cochrane as he exists in my mind." Still it was chiefly to Plutarch that Boswell, sweeping aside the cold and draped records of later biographers, returned; he had been deeply influenced by him in presenting the wholly Plutarchian figure of Paoli in the *Tour*. But to save Johnson from destruction required something more ponderous in mass and subtler in drawing. Large scale and effect in the great lines; but those lines to be animated, filled out, determined, by the thousand particles of which life truly consists. And the particles must be small, authentic and "characteristical." Boswell foresaw that his method might, in the eyes of his contemporaries, deprive his book of literary "dignity"; and in fact it was hailed as an amusing rather than a great work. But the propensity of his gift was urgent towards a plan which on reflection he perceived was, incidentally, "the best that can be conceived."

He was prompted to his method not only by a talent and a conviction; he was led to the same end by a cast of character. Recurrent in his diaries is a morbid horror of death and destruction which seems to have been a principal element in his hypochondria. Against "the end of the party" he retained all his life the passionate rebellion of a child. Grounded in old-fashioned piety by his mother and Mr. Dun, he could not banish the thought of the transitoriness of earthly pleasures. And, unhappily, this tragic quality of transiency was in nothing more marked than in what Boswell valued most—convivial happiness. At a London table the provincial Boswell found wit, wisdom, self-importance, drink, kindness, learning and "improvement": here, to use his favourite quotation from the *Beggar's Opera*, "every flower was united." But every time he set his face towards Scotland he reflected that, unless he could find means to garner it, the triumph and the laughter and the very words of Johnson were no more than a vanished parcel of air. His biographical labour was inspired by the same desperate resistance to the flux of things which caused him to preserve each most insignificant relic of his own life, in order to re-live it.

"*By how small a speck does the* Painter *give life to an Eye*," Boswell again observes, with his thoughts upon his own art. The small relic preserved is most potent to recall the past; the authentic word, the queer noted gesture of Johnson must be chosen and stored, in order to preserve and portray him "as he exists in my mind." The touch he aimed at was not to be the impressionist's summary stroke, but, rather, the carefully observed infinitesimal touch "in the Flemish picture which I give of my friend."

II

What steps did Boswell take to procure the fulness of immediate detail on which his scheme depended? This question has been variously guessed at; the Malahide papers furnish a fairly complete answer.

Johnson, on March 21, 1783, was expatiating on the different requisites of conversational excellence—Boswell, in the *Life* records, "While he went on talking triumphantly, I was fixed in admiration, and said to Mrs. Thrale, 'O, for short-hand to take this down!' 'You'll carry it all in your head; (said she;) a long head is as good as short-hand.' "

Boswell, if these words mean anything, felt *unable* to *"take down"* Johnson's talk.

But against this must be set a passage in the *Life*, dated 10 April 1778. "I this evening boasted, that although I did not write what is called stenography, or short-hand, in appropriated characters[1] devised for the purpose, I had a method of my own of writing half words, and leaving out some altogether so as yet to keep the substance and language of any discourse which I had heard so much in view, that I could give it very completely soon after I had taken it down." Challenged by Johnson to make the experiment, Boswell took down "part of Robertson's *History of America*," and "It was found that I had it very imperfectly."

To this must be added Mrs. Thrale's evident reference to Boswell when she writes of "a trick I have seen played on common occasions of sitting down steadily at the other end of the room to write at the moment what should be said either by Dr. Johnson, or to him." And this habit, she adds, "I never practiced myself nor approved of in another."

On the strength of these references several of Boswell's editors have accepted and enlarged upon the legend of Boswell as the reporter of Johnson, dashing down, as fast as he could, the words that fell from his lips. Croker is inclined to see in Boswell's shorthand and in his

---

[1] Boswell was, none the less, interested in stenography. His name figures in the list of subscribers to *The Complete Instructor of Short Hand* by W. I. Blanchard (undated, but, to judge by its decoration, apparently published about 1770). More than this, we find among his memoranda, as early as 1763, half a dozen scattered sentences written in stenographic symbols. These, however, do but confirm his statement that he did not use "appropriated characters" for speed: the signs are slowly and carefully formed. He uses them only as a cypher.

admission that he "had it very imperfectly" the explanation of some verbal errors and obscurities in his work. Fitzgerald, Boswell's biographer, says "he did not scruple to *report* regularly, and it would almost seem that he took so little share in what was going on, or was so privileged, that his proceedings caused as little *gêne* as a professional stenographer would to a practiced speaker."

Boswellians, I think, have always looked sceptically on the view which Croker here favours, and Fitzgerald firmly adopts. It attributes to Boswell a mean and insignificant role, fitted rather to Macaulay's caricature than to human probability. Professor Tinker, for one, has most explicitly repudiated the legendary view.[2] But Mrs. Thrale's attack, and Boswell's apparent admission, are "strong facts." Most writers have preferred to leave the question in a vague and crepuscular doubt; somewhere in the shadows around Johnson hangs Boswell with a notebook, "taking down" the talk in "a method of my own."

So far as the evidence goes, Mrs. Thrale's, though malicious in intention, cannot be set aside as untruthful. But, clearly, given her obvious desire to ridicule Boswell and some natural resentment at what she thought bad manners, her stab need imply no more than that Boswell did, on perhaps only two or three occasions, fall to scribbling notes in the presence of the company;—that he was actually, even so, attempting to "follow" Johnson is merely her guess. Boswell had retired and "sat down steadily" at the other end of the room and may very likely have wished to summarise some particularly valuable talk, immediately after listening to it. Boswell's own remarks will be considered in a moment, in the light of some of the Malahide papers.

But first it may be said that the legend of Boswell habitually, or frequently, attempting to "report" during the progress of conversation, can scarcely be reconciled with common sense.

(1) It is opposed to Boswell's character. We are asked to believe that he was willing "to take little share in what was going on." This is indeed difficult to picture. Boswell lived for conviviality; he was ambitious to shine. Above all he had the vanity—largely justified— of *steering* the moves of the conversation and drawing Johnson out.

(2) It is almost impossible to reconcile with eighteenth century manners. Johnson might have allowed it, for he was not unwilling to be recorded, and anxious to be accurately recorded. The rest of the group would be little likely to tolerate it. The "habit" imputed to Boswell would indeed in Mrs. Thrale's own words be "so ill bred . . . that were it commonly adopted, all confidence would soon be exiled from society and a conversation assembly-room would become tre-

[2] In *Young Boswell*, Chapter IX. The strongest argument for the traditional view —Boswell's statement to Johnson—is not there touched on. But Prof. Tinker's own balanced estimate is fully borne out by the new evidence of the Malahide MSS.

mendous as a court of justice." Yet there is plenty of evidence that Boswell was not only tolerated but welcomed. Of his diary and collections of *dicta* he made no secret; his importunity received occasional rebukes; but only after the *Tour* was published does Boswell begin to fancy that some people fight shy of him as a man who may write down their talk. In any case, if his attempts at stenography in public were anything but the most infrequent and insignificant, it would have been a standing joke, and we should have far more record of it than Mrs. Thrale has left us.

(3) It would have defeated Boswell's purpose. The best way of obtaining a living record, such as Boswell required of a conversation, is to listen to it critically; the very worst is to try, breathlessly, to get it down on paper. Dr. Burney[3] describes Boswell, in Johnson's presence, as indifferent to everything but the fear of missing "the smallest sound from that voice to which he paid such exclusive, though merited homage . . . The attention which it excited in Mr. Boswell amounted almost to pain. His eyes goggled with eagerness; he leant his ear almost on the shoulder of the Doctor; and his mouth dropt open to catch every syllable that might be uttered: nay, he seemed not only to dread losing a word, but to be anxious not to miss a breathing; as if hoping from it, latently, or mystically, some information." This is the picture of a man willing, for an enthusiasm, to make a fool of himself; intent on an enterprise greater than Dr. Burney guessed. It is, though a cruel picture, plausible and even convincing. But it is not the picture of a stenographer.

(4) It was evidently unnecessary. The *Life* abounds with conversations which took place under circumstances when such reporting was out of the question: *tête à tête* conversations—in a chaise, in a boat, on a seat in the sun. It is certain that these talks are not eked out by invention, for Boswell is scrupulous to say when he is unsure of a phrase or a word, even as between virtual synonyms. They cannot be the result of any reporting such as Mrs. Thrale describes. Yet these conversations are not inferior, either in fulness of detail or closeness of argument, to the others. Why, then, if he was equally able to dispense with it, should Boswell have employed on those other occasions an arduous art of immediate stenography which, in one statement, he confesses was very imperfect, and in another, he implies he completely lacked?

(5) In Boswell's newly found diaries there is no single reference to his having such a practice. Given the very extended range, chronologically, of these MSS, and his habit of dwelling on every point of his routine, this argument, though a negative one, is of strong force.

---

[3] Memoirs II, 194.

(6) It does not seem to have been observed that the interpretation commonly given to the essential passage in the *Life* is unsatisfactory to the point, almost, of absurdity. If Johnson had seen Boswell habitually, or repeatedly, *taking down* his talk he would not need *to be told* on April 10, 1778, *after fifteen years* of acquaintance, that Boswell had some method of shorthand. Nor is it thinkable that he would have waited all that time to put Boswell's capacity to the test. He would on the contrary, long before, and many times over, have been provoked to do so. We are asked to believe that after the material of *nearly four-fifths*[4] of Boswell's "Johnson" had been accumulated, Boswell, who according to Fitzgerald had during this time so often taken down Johnson's talk that his activity caused "as little *gêne* as a professional stenographer," suddenly boasts in 1778 that he possesses this gift. And Johnson, incredulous, defies him for the first time, and experiments, not with his own talk, but with Robertson's *History of America!*

III

What, then, is the true bearing of Boswell's own statement, already quoted, in the *Life*, about *"a method of my own"* of abbreviated writing, to preserve the substance of a conversation? For it is on this passage that the legend most solidly rests. The Malahide papers enable us to answer the question with some sufficiency. There are among them a great quantity of Notes often in dated sequence on uniform pieces of paper, sometimes in separate scraps. These Notes, which will be examined in a later section, are beyond any doubt what Boswell relied on for his reconstruction of Johnson's talk. *They are written, as Boswell describes, in his kind of shorthand;* ("writing half words, and leaving out some altogether") *and they are* not *written during the conversation, but (as will be shown in due course) when he gets home, at the end of the day, or next day.* They *account* for Boswell's remark, and do nothing to strengthen Mrs. Thrale's. Besides, what Boswell says is that his "method" enabled him "to keep in view" the substance and language of any discourse "which I *had* heard," (not "which I heard").

But, it will be objected, if the purpose was not to keep pace with the talk, why employ shorthand? The answer is that Boswell's original intention was to keep his full Journal up to date; that he found the daily labour repugnant; the Journal lagged behind and the precious *authentic contemporary* record was thereby jeopardised; he consequently formed the habit of *a rapid and highly condensed daily no-*

---

[4] I include in this calculation the *Tour to the Hebrides*, the Journal of which was in existence by the date in question.

tation, from which the events of his life and the full talk could (often many weeks afterwards) be "given very completely" in the Journal.

It is quite probable that Boswell first developed his boasted "method" (which after all is no different to what anyone uses who writes in a hurry) in the course of his work as a barrister: it is exactly what a man would use in taking down notes of evidence, etc. in court. In his summary of the Douglas case he claims that the greater part of the arguments are drawn genuinely from speeches of the Lords of Session, of which he had himself taken "very full notes." The habit of fast abbreviated writing, once formed, was employed by him to curtail the daily labour of his diary. And he was inclined to believe he *could* at need keep pace with actual talk. But that he relied on it as a method of recording Johnson, that it formed in any sense whatever the first basis of the *Life,* is, for the reasons given above, a conclusion at once so improbable and so unnecessary, that the legend should be discarded.

There remains this residuum. Boswell on some occasions, perhaps only one or two, at the Thrales', had the bad manners to draw aside from the company in order (as was understood) to note down what was being said—or, more probably, what had *shortly before* been said. And he may have sometimes done so elsewhere, when he was impatient to make an immediate record of what had already passed, or perhaps merely bored with what was actually passing. This picture is as likely as the other is unlikely. Such an occasional incident (sufficiently surprising in an eighteenth century society) would be grave enough to provoke Mrs. Thrale's reproof, and yet slight enough to account for our not hearing more about it from others. . . .

The "Notes"

The next class of the Malahide Papers to be examined is of high interest to students of the *Life of Johnson,* for without doubt we may recognise in it the original raw material of Boswell's book.

These Notes are not collections of Johnsoniana as such, but a condensed diary of Boswell's life. The Notes are nearly contemporary with the events. Johnson figures among the rest of Boswell's world, more fully and scrupulously preserved, but simply as one element among others in the record.

Of the genesis and purpose of these Notes, Boswell's Journal gives us ample information. On October 25, 1764 he writes: *"My method is to make a memorandum every night of what I have seen during the day. By this means I have my materials allways secured. Sometimes I am three, four, five days without journalising. When I have time and spirits, I bring up this my Journal as well as I can in the hasty manner in which I write it."* This was written when Boswell was on the Grand Tour; and he retained the habit here described to the end of his life. Like all diary-keepers, indeed, he dreamed of a Journal which should be "full" and always up to date. But more than most he fell short. Dissipation, laziness, hypochondria are excuses he alternately makes. But he had taken to heart Johnson's warning of the dangers of "dilatory notation" and every night, or next morning, he jotted down in "shorthand" a summary of the day's events and the day's conversation.[5] And this was done not as a permanent record, but simply to serve as a reminder when in a few days or weeks he found time to "bring up" his Journal. Now and again the ideal of a full Journal keeping pace with the events was realised for a few days, after the birth of a son, a New Year, an arrival in London, or at some other "fresh start"; and on these occasions he resolves never again to fall back. But he invariably reverts to the stopgap "shorthand" Notes.

The interval which was allowed to elapse between these and the writing up of the full narrative tended to lengthen from the "three, four, five days" mentioned in 1764 to a period of as many weeks. Boswell very frequently marks in his Journal not only the date of the events but the date of writing. An entry for 1 October, 1776 has the heading, *"Writing from* notes *(as I allways do)* 17 *October."* In the Ashbourne Journal, for September 24, Boswell notes, *"Writing at Auchinleck* 23 *October"*—a month's interval; and so on *passim*.

In these Notes there is little sign of any careful selection of Johnson's talk. It shows as a rule every mark of impatience and haste, and the amount set down varies from brief inchoate jottings to lengthy ill-scribbled reports. The impression we receive is rather that Boswell, without stopping to pick or choose, set down everything that rose to his mind in the brief time at his disposal. There is a parlour game in which a tray loaded with a bewildering variety of miscellaneous objects is set for a limited number of seconds before the eyes of the players who are then required in a further limited time to make a list of all they can remember. Many of Boswell's Notes have the air of just such a list. He is writing,—tired and perhaps a little fuddled,— at the end of a social day; and it matters little what he writes: he knows that any tag will suffice later on to bring the events and the talk back to his strong memory. For us, unless we chance to possess the clue, they are tantalisingly inadequate. This method, prompted originally by indolence, was probably the very best he could have chosen. After a brief interval a process of unconscious selection and

[5] Even the precious notes were sometimes neglected, shrinking to a mere skeleton list of dinners and suppers; and, now and then, *"I cannot remember where I dined."* The oblivion may charitably be taken as a proof of "dilatory notation."

artistic rendering would automatically begin to operate in his mind; yet the first "authentic" Notes were there to jog and control his memory. Reading over such notes as these, the whole scene was re-enacted at the right distance and, steeped in the Johnsonian "æther," he could recount what rose to his mind in the form of a lucid and dramatic narrative.

It was Boswell's practice systematically to discard his rough Notes as soon as he had expanded them into the fuller shape. Consequently, with rare exceptions, those which survive correspond to gaps in his diary; they relate to periods which he never found the energy to "write up." For this reason the proportion of Johnsonian Notes is now very small in comparison to those relating to the uneventful stretches of Boswell's Scottish life. But quite enough survive to enable us to form an exact estimate of their nature. There can be no doubt that almost all the *Life of Johnson*—of that essential part of which Boswell is a direct witness—was gathered in the first instance from records of this nature: from thousands of irregular pages similar to those here illustrated. It is to these that Boswell referred when he spoke to Johnson of "a method of my own of writing half words, and leaving out some altogether so as yet to keep the substance." They are Boswell's own life, waiting for incorporation in his Journals. These (and not any "shorthand" taken down from Johnson's lips) are the original bricks out of which the vast structure of the *Life of Johnson* was truly raised.

Papers Apart

We find among the Malahide MSS a small number of records which Boswell designates as "Papers Apart."

The distinction between these and his other MSS is merely that they fall outside the strict machinery of his Journal. The latter may be abandoned or, again, it may be complete with its daily series of entries. There will still be a number of loose sheets covered with memoranda of a similar kind.

I do not find that the Papers Apart are necessarily nearer to the events, or fuller in detail than the rough diary, though they commonly have these two characteristics. They tend to deal with an isolated occasion, and the dialogue is disentangled from irrelevant incidents.

Such papers owe their separate character, I think, merely to accidental circumstances. Sometimes even the intermediate short diary had fallen into arrears, and Boswell, without waiting to bring it up, writes out the gist of some conversation while it is still fresh in his memory. Or, conversely, he decides on second thoughts that an account already set down in the rough diary is an insufficient reminder, and rewrites it on a separate paper. . . .

The Journals

The next stage in Boswell's biographical process is represented by the Journals—the continuous narrative built up from the Notes, at an interval as a rule of from one to four weeks.

It is no exaggeration to say that Boswell regarded his Journal as the principal duty and aim of his existence; life unrecorded was not life. He goes so far as to make this singular pronouncement: *"I should live no more than I can record, as one should not have more corn growing than one can get in. There is a waste of good if it be not preserved."* Boswell did not feel he *possessed* an experience till it was written down: the *res gestæ* were mere preliminaries. *"True,"* he adds, *"the world would not hold the pictures of all the pretty women who have lived in it, and gladdened mankind; nor would it hold a register of all the agreable conversations which have passed,"* and so, if life serve the purpose of immediate felicity, *"perhaps that is enough."* Nevertheless in every act of life he is registering the scene, and analysing the sensation, with a view to putting it on record. As for disagreeable experiences, there is nothing, he says, that he cannot go through, *"if only I am to give an account of it."*

When his multifarious business gets behindhand, it is not the neglect of his law duties, his literary labours, his correspondence or his family obligations that gives him most concern, but always, with a peculiar sense of sin, *"my Journal."* *"Bring that up, and all will then be well,"* he reminds himself in the "memoranda" which formed a separate file in the vast book-keeping operation of his life. The Journal seems to have represented a kind of sheet-anchor in Boswell's veering existence. To set down those wavering purposes, those fitful and inconsistent actions and then to contemplate them on paper, was the nearest approach to a sense of stability he could achieve.

What is very remarkable, and perhaps surprising, is that Boswell's Journals (after 1765) are in the fullest sense of the word Private. The early Scottish Journal of 1762 (see Vol. I) and the Continental Journal are often self-conscious; he clearly anticipates that they will be handed round amongst his friends. But after his return to England all trace of deliberate art disappears. The itch for publicity, so marked in his nature, does not here guide the turn of a single phrase, distort a motive, or modify an incident. He is writing for himself. The Journals are a great reservoir of experiences and sensations, and Boswell, with his insatiable interest in his ego, has no thought but to make it scientifically true. He is well aware, and often reminds himself, that from this reservoir he can and may, at leisure, draw off material of numberless articles, "accounts" and even books: above all, a *Life of Johnson.* But the future publication of the Journal, or of any considerable part of it *as it stands,* is never in his thoughts. There is no attempt at "effect." The histrionic talent of Boswell, irrepressible when he has an audience even of one, and so evident in his letters, is here in abeyance. But what is thus lost in art is more than gained in authenticity. Candour can go no further. The Journal does not exist to show Boswell to advantage, but to objectify him in his own eyes.

The *Life of Johnson,* in the years when Boswell knew him, is essentially a part of this Journal, sharing the sincerity which marks the whole record. It is a part which is, of course, from the very first conceived as *possible* material for a literary use; but it is not (at the unrevised stage) consciously literary. The Johnson-record flows in and out of the personal Boswell-record and is not different in kind. The vast, bracing difference is the subject matter.

It no doubt contributed to the strength of the final result that the first full record should have been of this private and unstudied character. Had Boswell composed in the first instance with an audience directly in view, he might probably not have avoided occasional lapses into the declamatory style which so often inflates his letters, published articles and pamphlets. He might have puffed his subject. One cannot tell; perhaps his sense of reverent discipleship towards Johnson might, even so, have sufficiently protected him. But it is certain that in its simplicity and absence of rhetoric the *Life of Johnson* resembles the private unliterary Journals rather than the other published works. It was written *in the first instance* by Boswell the observer and not by Boswell the showman. And if anything can increase our sense of the ascetic veracity of the *Life of Johnson* it is to read the narrative, as most of it was first written, mingled with the patient tale of Boswell's acts and hopes and humiliations, set down for his own solitary view.

A large proportion of Boswell's London Journals would seem to have been destroyed in the process of compiling the *Life.*[6] But considerable sections of the MS are still intact, and are amply sufficient to establish the general relation between the contemporary and final record. They will be printed in due course. Here, to complete our general survey of the successive stages of the biographer's process, the "Journal" is illustrated by a *dossier* of papers which carries Boswell's endorsement, *"Journal in London,* 1773."

It consists (1) of fifty-two sewn but unbound pages describing Boswell's journey from Edinburgh and the earlier part of his visit to

---

[6] [Later discoveries proved this to be a wrong conclusion.—ED.]

London in the spring of that year, (March 30 to April 13). Twenty-six of these pages were "written up" at one sitting on April 30th, (*See* his Note for that day, page 130). (2) From April 11th till his return to Edinburgh we find a rough untidy record on loose quarto and folio pages. They form a continuous daily diary, supplemented by three "Papers Apart." Their careful preservation, in the same endorsed wrapper with the earlier "full" Journal, makes it almost certain that no formal Journal for these later weeks was ever made up. And it is easy to understand, when we examine the lengthier dialogues, that Boswell should have been tempted to dispense with the labour of an expanded transcription. It will be found that the narrative in the *Life* for the spring of 1773 is all but fully accounted for by these documents. And I do not doubt that they are the precise and (with insignificant exceptions) the *only* MSS from which he eventually worked.

The Caldwell Minute

by Frank Taylor

As a sample of how Boswell used an important original source when fashioning one of his best known scenes, Johnson's interview with the King in 1767, a recently discovered document is of great value. Reputedly written down shortly after the event, this is Johnson's own version, a copy of which was sent to Sir James Caldwell, and which is now in the John Rylands Library in Manchester, England. It was first printed in 1952, together with a detailed description and analysis by Dr. Frank Taylor, the Rylands Keeper of Manuscripts. Because of limitations of space, only the "minute" itself and Taylor's final discussion are included here. Omitted are physical descriptions of the manuscript, his analysis of the various emendations of the text, annotation of individual references, speculations as to whether this was an original dictated text or a later copy (except for a few possible late additions, it is not in Johnson's handwriting), and other detailed matter. For any thorough study of the "minute," Taylor's article must be consulted— ED.

'The King came in and after having ~~walked by Mr. Johns~~ talked for some time to the other persons in the Library, turned to D^r. Johnson and asked him if he were not lately come from Oxford. D^r. J. answered that he was, upon which the K. again asked ~~how~~ him if he were not fond ~~of Oxford~~ going to Oxford to which he replied that he was indeed fond of going to Oxford but was likewise glad to come back again.

The King then asked him if he were writing anything at present to which he made answer that ∧ he was not for he had pretty well told the world what he knew, and ~~that he~~ must now go and read for more. I do not think D^r. J. replied the King that you borrow from any one, upon which D^r. J. observed that he thought he had pretty well done his Share. I should think so too D^r. J. said the K. if you had not done so well.

"The Caldwell Minute" (editor's title) by Frank Taylor. From "Johnsoniana from the Bagshawe Muniments in the John Rylands Library: Sir James Caldwell, Dr. Hawkesworth, Dr. Johnson, and Boswell's Use of the 'Caldwell Minute,'" Bulletin of the John Rylands Library, XXXV (1952–53), 235–47. *Reprinted with permission from the* Bulletin of the John Rylands Library.

The K. then proceeded to ask him what they were doing at Oxford. Upon which M^r. J. told him that ~~they~~ he could not indeed much commend their diligence but that in some respects they were mended for ~~that~~ Press they had put their under better regulations, and were at this time printing Polybius. He was then asked whether they had better Libraries at Oxford or Cambridge, to this he replied that he believed the Boldleian was larger, ~~at the same time observing~~ than any they had at Cambridge, but at the same time added that he hoped ~~wheth~~ whether we had more books or not, that we should make as good use of them as they did. He was then asked whether All Souls or Christ Church Library were the larger, to which he [f.lv.] replied that All Souls Library was the largest we had except the Bodleian, Aye said the K. that is the Publick Library. The K. then told him that he thought he must have read a vast deal. D^r. J. replied that he had thought more than he had read, that he had not indeed neglected reading, but having very early in life fallen into ill health he had not been able to read much compared with those who had, for instance he said he had not read much compared with D^r. Warburton. Upon which the K. said that he had heard D^r. W. was a man of very general knowledge, and that you could scarce talk with him on any subject upon which he was not qualified to speak. The K. then asked him what he thought of the Controversy between D^r. W. and D^r. L. To this he replied that he thought D^r. L. called names rather better than D^r. W. You do not think then said the King D^r. J. that there was much argument in the Case. He said he did not think there was. Why true said the K. when once it comes to calling names ~~all~~ argument is pretty well ~~over~~ at an end. He then asked him what he thought of Lord Lyttelton's book just published. He said he thought his Style might be pretty good [but that he had blamed Rich ʒ. ~~by wholesale~~ rather too much]. Why says ye K. they seldom do these things by halves. No Sir said he not to Kings. But fearing that he might be misunderstood he ~~the~~ proceeded to explain himself and immediately subjoined that for those who spoke worse of Kings than they deserved he could offer no excuse, but that he could easily conceive ~~how~~ how [f. 2] some might speak better of them than they deserved ~~th~~ without any ill intention, for as Kings had much in their power to give, those who were much obliged to them would frequently from Gratitude exaggerate their praises, and as this proceeded from a good motive it was certainly excuseable as far as error could be excusable. He then asked what he thought of D^r. Hill, when he told him that ~~he thought~~ he was an ingenious man but had no verac-
as an instance of it
ity, and immediately produced ∧ that assertion of D^r. Hill's viz. that he
seen things
had ∧ magnified to a much greater degree by ~~looki~~ using ʒ or 4 microscopes at a time, now said he every one who has seen a microscope knows that the more he looks thro' the less he will see. Why replied the K. this is not only telling a falsehood but telling it clumsily for if this be the case, every one who can look through a microscope will be able to detect him. D^r. J. then proceeded to tell the K. that D^r. H. was notwithstanding a very curious observer and if he would have been con-

                                                                    cut
tented to tell the world no more than he knew, he might have ~~made~~ a
very considerable figure in the world, ~~with~~ and needed not to have
recourse to such mean expedients to raise his reputation. They then began to speak of Literary Journals the nature and use of which Dᴿ. J.
explained, and at the same time gave some account of the principal
writers on that subject, he told who first begun the Journal des Scavans
and said that it had been well written for many years, upon which the
King asked if it were well written [f. 2ᵛ.] now? he said he had no reason
to think that it was. The K. then asked him the character of our two
                        [were]
Reviews and whether there ∧ any other literary Journals now published
in ~~England~~ this Kingdom, he said there were no Literary Journals be-
                                      Critical
side these, and that of these two the ~~Monthly~~ was written with the least
                                            that
care and the Monthly on the worst principles. For the authors of the
                                this
monthly Rev. were enemies to the Church, ~~which~~ the K. said he was
sorry to hear. The Conversation next turned on the Philosophical Trans-
                                          had
actions when Mᴿ. J. observed that they ~~had~~ now a better method of
sorting their materials, Aye said the K. they are obliged to Dᴿ. J. for that.
       For
~~This Circumstance~~ his Majesty had heard and remembered this Circumstance which Dᴿ. J. had himself forgot. This was I bllieve all ~~the Conver-~~
~~sation~~ that passed between them, for about this time ~~the Princess Dow-~~
~~ager came in and put an~~ a visit from the Princess Dowager put an end
to the Conversation.'

The Caldwell Minute is apparently the only one of Boswell's five
sources for this incident which has survived and it is fortunate that it
should prove to be the longest, providing him with more material
than all the rest together. But in spite of its length, it does not add
any new information to that already known; Boswell's own interest
in the royal interview was such that it would be surprising if it did.
Its value consists rather in this: that it enables us, firstly, to follow
through its various stages the evolution of the major portion of a
passage which he himself considered one of the most important in the
Life, and at the same time test the accuracy with which he handled a
written source, and, secondly, to analyse his whole account of the
royal interview.

The main stages which may be recognized in considering Boswell's
handling of this source are: the Minute itself; the copy of it which,
he informs us in the *Life,* was obtained for him from Caldwell's son,
Sir John, by Sir Francis Lumm; his own holograph account of the
whole incident put together ready for inclusion in the main manuscript of the *Life* and now preserved at Yale among his *Papers apart;*[1]

[1] The full reference to this document (cited here as *Paper apart* 320) is "Private
Papers of James Boswell, Yale University Library, MS. of *Life,* Paper apart, p. 320."

his proofs; and, finally, the version which appears in the published
Life. The differences between the third stage and the final version are
negligible, so that we may concentrate on the first three. Of these the
second is missing; but its loss is not serious for, as will appear from
what follows, it was a fairly close copy of the Minute. It is, in short,
only necessary to compare the Minute with the Yale manuscript in
order to discover the relation of the original material to Boswell's
finished work and to separate him from his source. For all his major
modifications had already been introduced into the Yale manuscript,
and introduced in such a way that they may clearly be picked out.
His method, with one exception, referred to below, was simply to
copy out the text of the Minute and then go over it making such verbal
alterations as seemed to him to improve its effect; accordingly, in the
Yale manuscript we see him actually at work on the text of the
Minute, crossing out, interlineating, introducing *oratio recta*, changing his mind on various points, changing it again, and then perhaps
deciding to retain the original wording after all. The exception is
the passage dealing with literary journals, where Boswell uses the
same material as that provided by the Minute, but for the most part
does not follow its wording; the reason for this departure is not clear,
but it may, to some extent at least, be due to a desire not to cause
offence. His different treatment of this passage, however, only serves
to emphasize his normal practice for, as far as the rest is concerned,
whether his subsequent alterations are numerous or nil, he always
begins with the text of the Minute, not his own version of it or even
a selection from it. It is indeed only necessary to ignore his alterations
in the Yale manuscript in order to recover the text of the Minute,
apart from a few insignificant differences which may be due to the
intervention of the copy he used. [Examples of alterations here
omitted—ED.]

Finally, the Minute assists us in analysing his whole account of the
royal interview. His narrative, we are told in the *Life*,[2] was compiled
from five sources: "from Dr. Johnson's own detail to myself; from Mr.
Langton, who was present when he gave an account of it to Dr. Joseph
Warton, and several other friends, at Sir Joshua Reynolds's; from
Mr. Barnard; from the copy of a letter written by the late Mr. Strahan
the printer, to Bishop Warburton"; and from the Caldwell Minute.
But although the names and number of Boswell's sources have been
known, the extent to which he relied on each has not; nor does he
even mention them all in the course of his narrative. But given the
Minute we may by a process of elimination discover the rest, for,
that removed, it can be seen that what remains contains for the most
part its own indications of source in the form of words such as "John-

[2] *Life*, ii. 34, n. 1.

son observed to me," [3] "at Sir Joshua Reynolds's," [4] "said Johnson to his friends," [5] "he said to Mr. Barnard," [6] and "he afterwards observed to Mr. Langton." [7] Built into and around the Minute, his main source, and previously obscured by it, the passages dependent on these indications can now be clearly recognized. Together with the Minute they account for all his sources save one, Strahan's letter, and cover the whole of his account of the royal interview save ten lines. Of these ten, four[8] deal with the King's suggestion that Johnson should write the literary biography of this country; their origin is not clear, unless, like the passage immediately following, they are from Barnard. The remaining lines occur in the discussion concerning Warburton and Lowth and it seems probable that these came from Strahan's letter to Warburton, which, apart from being a likely source, is also the only one not yet accounted for.

It may be noted in conclusion that Boswell originally intended to include more material from Strahan's letter. As his *Paper apart* 320 shows, his entire account of the discussion on historians and Kings was first taken from it. But he changed his mind, struck it through and marked it "see oppos[i]te," a reference to the verso of the preceding leaf where, under the heading "Caldwell," he copied out the corresponding passage from the Caldwell Minute which replaces it both there and in the version he finally printed in the *Life*.

[3] *Ibid.* ii. 35, l. 27.
[4] *Ibid. ii.* 35, l. 30; 41, l. 4; 42, l. 12.
[5] *Ibid.* ii. 39, l. 6.
[6] *Ibid. ii.* 40, ll. 21–22. The end of the Minute corresponds to l. 11 of this page.
[7] *Ibid.* ii. 40, l. 23.
[8] *Ibid.* ii. 40, ll. 12–15.

Boswell's Johnson

by Marshall Waingrow

Geoffrey Scott, in concluding his study of the making of the *Life,* preferred the conclusion in which not everything is concluded. His argument is all but irresistible:

> To speak of the final achievement to which all this labour was directed, would perhaps be a logical, but certainly an impertinent, conclusion. It is a quality of the *Life of Johnson* that its appeal is not less intimate than universal; the praise or comment of a critic, even when we agree with his statement, falls like a hand officiously and needlessly intruded on a private possession. This silencing consideration, if it applies in regard to every reader of the *Life,* has still greater force in the presence of the Johnsonian scholars for whose convenience these documents have been assembled, and who may, with better right, fulfill what is not here attempted.

Like Scott, we have pushed to our own logical conclusion (indeed, the cases are virtually identical), and if we cannot, despite his example, resist speaking of the final achievement, we can only hope that the impertinence will be forgiven by common readers and scholars alike.

This introduction began with the suggestion that the greatness of Boswell's biography owes as much to its wholeness as its fullness, as much to the synthesis of its multitudinous facts as the facts themselves. Any analysis of Boswell's editorial practices must be tested finally against a conception of the finished work; the limited contexts in which we view the particulars of Boswell's method must be enlarged to include the spectacle of the whole.

Admittedly, the spectacle of the whole of the *Life of Johnson* is bewildering. The disjointed and unbalanced narrative deprives the reader of one form of regularity upon which the eye might rest. More-

---

"Boswell's Johnson" by Marshall Waingrow. From The Correspondence and Other Papers of James Boswell Relating to the Making of the Life of Johnson, *ed., Marshall Waingrow (New York: McGraw-Hill Book Co.; London: William Heine-mann Ltd., 1969), pp. xliv–l. Reprinted with permission of the McGraw-Hill Book Company, William Heinemann, Ltd., and the author.*

over, though Boswell's scheme is chronological, it is in no real sense
progressive. Instead of the growth of a mind (the absence of which
modern critics have repeatedly deplored), Boswell deliberately shows
us its unaltered passage through the world. In this light neither completeness nor order matters so much as singleness of impression. Yet
Boswell's tireless marking of "the most minute particulars" in his
"Flemish picture" conveys an extraordinary variety—or at least the
illusion of it. The range of the conversations is probably the principal
tributary to this effect; but the conversations themselves illustrate,
what Boswell's own narrative so often makes explicit, a certain level
of uniformity in Johnson's life.

No trait of Johnson's receives more emphasis in the biography than
his intellectual powers and the use to which he put them:

> But his superiority over other learned men consisted chiefly in what
> may be called the art of thinking, the art of using his mind; a certain
> continual power of seizing the useful substance of all that he knew,
> and exhibiting it in a clear and forcible manner; so that knowledge,
> which we often see to be no better than lumber in men of dull under-
> standing, was, in him, true, evident, and actual wisdom (iv. 427–28).

Johnson's pre-eminence of mind is insisted upon throughout. He had
"a knowledge of Latin, in which, I believe, he was exceeded by no
man of his time" (i. 45). He was the intellectual superior of his fellow-
students and indeed of his teachers. In the development of English
prose "he appeared to lead the national taste" (i. 222). "In biography
there can be no question that he excelled, beyond all who have attempted that species of composition" (i. 256). In the writing of dedica-
tions "no man excelled Dr. Johnson" (ii. 1). Johnson enters upon a
company and "we were all as quiet as a school upon the entrance of
the head-master" (iii. 332).

Goldsmith complained to Boswell "for talking of Johnson as en-
titled to the honour of unquestionable superiority. 'Sir, (said he,) you
are for making a monarchy of what should be a republick'" (ii. 257).
The metaphor is in fact Boswell's own favourite way of expressing Johnson's supremacy.[1] We have already noted [see *Life*, i, 47–48] the passage
in which Johnson the schoolboy, "a king of men," was "borne trium-
phant." He was "a majestik teacher of moral and religious wisdom"
(i. 201). "His majestick expression would have carried down to the
latest posterity the glorious achievements of his country" (i. 355). Upon
Johnson's writings Boswell reflects: "Tastes may differ as to the violin,
the flute, the hautboy, in short, all the lesser instruments: but who
can be insensible to the powerful impressions of the majestick organ?"

---

[1] Compare Courtenay's *Moral and Literary Character of Dr. Johnson:* "By nature's
gifts ordain'd mankind to rule" and "his philosophick throne" (*Life* i. 222–23).

(ii. 335). But the best illustration of this view of Johnson is a dramatic one: the famous interview with the King. For the ironic effect of the episode is to establish the majesty of Johnson, not of George III. It is to Johnson's authority that all questions are put; his also the privilege of stooping to compliment. He "talked to his Majesty with profound respect, but still in his firm manly manner, with a sonorous voice, and never in that subdued tone which is commonly used at the levee and in the drawing-room. After the King withdrew, Johnson shewed himself highly pleased with his Majesty's conversation and gracious behaviour" (ii. 40).[2]

Goldsmith's complaint is essentially that of all those who, without Goldsmith's personal grievance, charge Boswell with undue veneration of Johnson. Reverence is undoubtedly Boswell's last word on the subject—and indeed the last word of the *Life*. But the attitude of reverence calls for examination before calling for apology (as Carlyle saw). The royal metaphor points to the particular meaning of Boswell's awe: what he worshipped was not the mind for its own sake, but its power to *govern*. That there was a deep private need underlying Boswell's devotion is clear enough:

> I complained of a wretched changefulness, so that I could not preserve, for any long continuance, the same views of any thing. It was most comfortable to me to experience, in Dr. Johnson's company, a relief from this uneasiness. His steady vigorous mind held firm before me those objects which my own feeble and tremulous imagination frequently presented, in such a wavering state, that my reason could not judge well of them (iii. 193).

In paying this tribute Boswell lightly passes over Johnson's admission in the preceding paragraph that he was *not* always "the same." The infirmity of Johnson's mind (revealed so strikingly in the diaries) was a fact that his biographer recognized; but the greater fact for him was Johnson's constancy in spite of it.

"To have the management of the mind is a great art", says Johnson (*Life* ii. 440). The mind in this maxim is both object and subject, and the Johnsonian triumph, according to Boswell, is the triumph not

---

[2] Boswell acknowledges no fewer than five different sources for this narrative, any or all of which may have contributed the suggestion of a reversal of stations. A comparison between Boswell's version and its principal source, the "Caldwell Minute," shows Boswell setting the stage for his drama by sweeping some of the *dramatis personae* off it: namely, other persons present in the library, to whom the King "talked for some time" before turning to Johnson (F. Taylor, *Johnsoniana from the Bagshawe Muniments in the John Rylands Library: Sir James Caldwell, Dr. Hawkesworth, Dr. Johnson, and Boswell's Use of the "Caldwell Minute"*: reprinted from *Bull. Rylands Lib.*, 1952, xxxv. 211–47).

only of mind over matter (poverty, neglect, disease) but of mind over
mind itself (the dangerous prevalence of imagination).

> We cannot but admire his spirit when we know, that amidst a complica-
> tion of bodily and mental distress, he was still animated with the desire
> of intellectual improvement (ii. 263).

> Notwithstanding his afflicted state of body and mind this year, the fol-
> lowing correspondence affords a proof . . . of . . . his extraordinary
> command of clear and forcible expression (iv. 149).

> In 1783, he was more severely afflicted than ever . . . but still the same
> ardour for literature, the same constant piety, the same kindness for his
> friends, and the same vivacity, both in conversation and writing, distin-
> guished him (iv. 163).

Johnson's weaknesses, as our survey of Boswell's sources disclosed,
are methodically viewed under the aspect of his strengths: his indolence
together with his energy, his excesses of appetite together with his
abstemiousness, even his sexual irregularities together with the force
of his conscience (*Life* iv. 395–98). But it was Johnson's "morbid
melancholy" that fascinated Boswell most and provided him with his
major theme. If the mind is the most powerful weapon for coping with
the world, it is also the most vulnerable target for the assailants of
life. Johnson's greatest strength and his greatest weakness were near
allied: the mind that preys on everything will at times prey upon
itself. As we have seen, Boswell emphatically refused the allegation
of insanity in Johnson and instead presented him as ever rising above
his affliction. One early episode dramatizes the achievement in a
striking way: Johnson's diagnosing his own melancholy for his own
physician, Dr. Swinfen, in 1729. Boswell comments:

> The powers of his great mind might be troubled, and their full exercise
> suspended at times; but the mind itself was ever entire. As a proof of
> this, it is only necessary to consider, that, when he was at the very worst,
> he composed that state of his own case, which shewed an uncommon
> vigour, not only of fancy and taste, but of judgement (i. 65).

The worst is not, when we can say the worst; and a Johnson **can** push
the paradox further by saying it best.

Saying it best describes Johnson's talk as well as his writings, and
his oral prowess may be seen as both a symptom of the radical cause
of his melancholy and its palliative. We know that society was for
Johnson an escape from the horrors of solitude, but (despite the self-
deprecating self-portrait of Mr. Sober in *Idler* 31) talk meant more
than distraction: it was action itself. Johnson's conversation "will best
display his character," says Boswell (*Life* i. 31); and indeed, as anxiously preserved, or rather recreated, by Boswell, it is the most dramatic expression in the biography of the theme we have been tracing.

The image of Johnson as dogmatist probably owes more to the records of his talk than to his writings, for it is in conversation that he is most opinionated, peremptory, and violent. Yet it is a peculiarity of Johnson's mode of talking that it was deliberately defensive; Boswell tells us that he rarely initiated conversation, which, we may imagine, was for him a microcosm of the world: a flux of sentiments and beliefs, unsettling to a greater or lesser degree, and therefore requiring to be answered. For Johnson was dedicated, nay addicted, to settling notions. "Oglethorpe, Sir, never *completes* what he has to say." "Sir, there is nothing *conclusive* in [Lord Elibank's] talk." "Goldsmith had no settled notions upon any subject; so he always talked at random" (*Life* iii. 56–57, 352). To settle the most minute question was to affirm the authority of the mind, which meant in effect to put the world temporarily back in order. It is only before the great question of the greater order of the "other" world, the question of "futurity," that the mind of Johnson shrank. "He talked to me upon this awful and delicate question in a gentle tone, and as if afraid to be decisive" (*Life* iii. 200; cf. iii. 154, iv. 177). This is the Johnson of private life, of the diaries and prayers—"unsettled and perplexed," as Boswell describes him (*Life* iii. 98). But the Johnson of the *Life* is combative and, though not all-conquering, at least impressively holding his own:

> His mind resembled the vast amphitheatre, the Colisaeum at Rome. In the centre stood his judgment, which, like a mighty gladiator, combated those apprehensions that, like the wild beasts of the *Arena,* were all around in cells, ready to be let out upon him. After a conflict, he drove them back into their dens; but not killing them, they were still assailing him (ii. 106).

Boswell's simile is meant to represent Johnson's fear of death specifically, but it serves as well as an image of the whole tragic striving of his life. For Johnson is surely an eminent case of man girding his constancy before his inevitable change, and presuming to make definitive pronouncements in a finite world. Boswell, who knew how perilous this balance was, could still regard Johnson as *semper idem,* the apotheosis of sanity.

What it took to create such a portrait is pithily suggested in a passage joining in characteristic fashion a Johnsonian saying and a Boswellian reflection:

> Johnson said, "A madman loves to be with people whom he fears; not as a dog fears the lash; but of whom a person stands in awe." I was struck with the justice of this observation. To be with those of whom a person, whose mind is wavering and dejected, stands in awe, represses and composes an uneasy tumult of spirits, and consoles himself with the contemplation of something steady, and at least comparatively great (iii. 176).

Let us do for Boswell what he did for Johnson, and allow his own melancholy to be free of the implication of madness. Yet, in missing the irony of his analogy, did he not hit the mark of his achievement? If Johnson alive was that something steady, and steadying, of which Boswell stood in awesome contemplation, the writing of the *Life* was more than memorial therapy, the patient ministering to himself; the hypochondriac turned artist steadied his own doctor and consoled *us* with the contemplation of something at least comparatively great.

Our judgement of Boswell the editor must ultimately rest, we have argued, upon our judgement of Boswell the biographer—upon, that is, an appreciation of the finished work of art. If the fullness of that work has been more generally remarked than its wholeness, a study of Boswell's sources should subtract something from the one quality and add it to the other. And if the result does nothing to alter the long and widely held view that Boswell's Johnson is an "idealized" portrait, it may help alleviate our suspicion that what is idealized is therefore of necessity untrue. That the *Life* might have contained more truths and fewer errors is of course obvious; and it has been the work of scholarly commentary to bring it progressively to that desired condition. But it is equally obvious that, no matter how many new facts are brought to light, Samuel Johnson will always be somebody's hypothesis. And none has pleased so many, or is likely to please so long, as Boswell's.

Boswell's Control of Aesthetic Distance

by Paul K. Alkon

I

Proper control of aesthetic distance was so highly regarded by Johnson that he was sometimes inclined to undervalue biography. Thus in the *Idler*, No. 84, he argues that autobiography is more useful because "he that recounts the life of another, commonly dwells most upon conspicuous events, lessens the familiarity of his tale to increase its dignity, *shews his favourite at a distance* decorated and magnified like the ancient actors in their tragick dress, and endeavours to hide the man that he may produce a hero." [1] Hence the failure of most biographers. They keep their heroes too far away from us while, paradoxically, making them seem larger than life-size. Johnson's ideal for life-writing is clear: the less distance between reader and subject the better. Equally clear is Boswell's conscious adherence to that ideal.

Indeed Boswell's fame as an instigator of modern biography rests largely on his thorough rejection of the "doctrine of dignified distance." [2] Using a variety of devices which are well recognized by critics, Boswell succeeded in bringing his readers close, often uncomfortably close, to Johnson. Early in the *Life of Johnson* and only four paragraphs after referring to the argument in the *Idler*, No. 84, Boswell explains his decision to let as little as possible, especially of the narrator, stand between readers and Johnson. . . . Neither Boswell nor his critics, however, have pointed out the crucial devices employed throughout the *Life* to increase and, in general, vary aesthetic distance in order to solve some of the literary problems confronting the biographer. . . .

"Boswell's Control of Aesthetic Distance" by Paul K. Alkon. From University of Toronto Quarterly, *XXXVIII (January 1969), 174–91. Abridged and reprinted by permission of the author and the publisher, University of Toronto Press.*

[1] Samuel Johnson, *The Idler and the Adventurer*, ed. W. J. Bate, John M. Bullitt, L. F. Powell (New Haven 1963), 262. Italics added.

[2] Joseph W. Reed, Jr., *English Biography in the Early Nineteenth Century* (New Haven 1966), 38–41.

As the minute particulars pile up, through hundreds of pages, there is the problem of preventing readers kept this closely in Johnson's company from becoming so used to him that they forget what a remarkably *rara avis* Boswell is keeping in his biographical cage. Wayne Booth has correctly observed that "a prolonged intimate view of a character works against our capacity for judgment." [3] Any judgement, he might have added, whether of merit or merely of singularity. There is thus considerable danger that our very familiarity with Johnson, induced by such close acquaintance with "what he privately wrote and said, and thought," may induce us to lose track of his astonishing uniqueness. Yet for the *Life* to succeed readers must at the conclusion still be able to feel the force of Hamilton's moving farewell to his friend: "He has made a chasm, which not only nothing can fill up, but which nothing has a tendency to fill up.—Johnson is dead.—Let us go to the next best:—there is nobody;—no man can be said to put you in mind of Johnson." (IV, 420–21)

II

Not all of Boswell's artistic problems in writing the *Life* were dealt with entirely or even partly through control of aesthetic distance, to be sure. Most notably, his success in maintaining a coherent image of Johnson's character as a unifying principle of the *Life* was achieved by other means and therefore lies outside the scope of my discussion. . . .[4]

Since Boswell's professed goal is to make readers "live o'er each *scene*" with Johnson, the *Life* is committed to the methods of drama. And to describe a performance as "dramatic" was then as it still is a way of saying that it is interesting. Going beyond the metaphor, however, critics are now in agreement on how, in general, the *Life* succeeds in aspiring to the condition of drama. There are stage directions: "Johnson (smiling), Sir. . . ." There is dialogue. There are even some conspicuous episodes such as the Wilkes dinner which are given the beginning-middle-and-end structure of a well constructed play.[5] In many of the more dramatic episodes, moreover, Boswell as narrator-dramatist is appropriately out of sight behind the scenes: having set the stage, he minimizes the distance between audience and events by cutting down references to himself ("I kept myself snug and silent")

[3] Booth, *The Rhetoric of Fiction*, 322.
[4] See Ralph W. Rader, "Literary Form in Factual Narrative: the Example of Boswell's *Johnson*," *Essays in Eighteenth-Century Biography*, ed. Philip B. Daghlian (Bloomington 1968), pp. 3–42.
[5] Sven Eric Molin, "Boswell's Account of the Johnson-Wilkes Meeting," *SEL*, III (Summer 1963), 307–22; Bronson, *Johnson Agonistes*, 77; Frederick A. Pottle, "Boswell Revalued," *Literary Views*, ed. Carroll Camden (Chicago 1964), 79–91.

so that attention is focussed on the other actors surrounding his hero.[6] And because the essence of drama is talk, it is tempting to add to our growing list of critical commonplaces about Boswell's dramatic technique the fact that his commitment to dramatic method dictated a simple principle of decorum by which relevancy could be separated from tedious digression: commenting on his decision to exclude some "pleasant conversation" that Johnson had one day enjoyed hearing but in which he had not taken part, Boswell asserts that Johnson's "conversation alone, or what led to it, or was interwoven with it, is the business of this work." (II, 241–42)

But this plausible-sounding assertion will hardly do as an accurate or sufficient account of Boswell's method even at its most dramatic. In fact, the *Life's* ability to sustain interest is due largely to Boswell's willingness to violate every aspect of the principle of decorum he so sweepingly enunciates here. He often includes material that is not part of Johnson's conversation or his life, that did not occasion Johnson's remarks, and that was in no direct sense "interwoven" with them. But this is not to say that such material is unrelated to Boswell's subject. Rather, it is to suggest that the relationship is far different from that which Boswell claims in his explicit statement of what may properly find a place in his book. That remark more accurately describes the effect than the methods of his artistry: where the *Life* is successfully dramatic we are often only made to feel that Boswell has given us exclusively Johnson's talk, its causes, and what *"was"*—at the time the scene took place—"interwoven" with it. Sometimes we are indeed given these things. Often, however, the feeling is dramatic illusion. We have been induced to willing suspension of distinctions between past and present, as well as to suspension of our awareness of the difference between action on-stage and action off-stage.

Consider, for example, the following paragraph, complete in itself, and taken from a part of the record for 1776 where Boswell says that "to avoid a tedious minuteness" he will "group together what I have preserved of his conversation during this period . . . without specifying each scene where it passed" since "where the place or the persons do not contribute to the zest of the conversation, it is unnecessary to encumber my page with mentioning them." (III, 52) The dramatic method has been modified to the extent of dropping stage directions and the list of dramatis personae involved, but only in order—Boswell claims—to render the conversation, still his professed subject, as vigorously as possible:

> "There is much talk of the misery which we cause to the brute creation; but they are recompensed by existence. If they were not useful to man,

[6] Molin, "Boswell's Account of the Johnson-Wilkes Meeting," 320–21.

and therefore protected by him, they would not be nearly so numerous."
This argument is to be found in the able and benignant Hutchinson's
'Moral Philosophy.' But the question is, whether the animals who endure
such sufferings of various kinds, for the service and entertainment of
man, would accept of existence upon the terms on which they have it.
Madame Sévigné, who, though she had many enjoyments, felt with
delicate sensibility the prevalence of misery, complains of the task of
existence having been imposed upon her without her consent. (III, 53)

What Johnson actually said occupies only the first two sentences, less
than half of the passage. His opinion is followed by the seemingly
digressive and gratuitous information that Johnson's opinion was also
held by the Scot, Hutcheson. Conspicuously omitted is any claim that
Johnson was influenced by *Moral Philosophy*. Indeed so far as Boswell
knew, or at least so far as he reports in the *Life,* Johnson had not even
read Hutcheson's book. Instead of urging any relationship other than
coincidence of opinion between the two moralists, Boswell chooses to
praise *Hutcheson's* ability and benevolence. Boswell as narrator then
moves to the front of the stage where he proceeds in the next sentence
to soliloquize on what the question is: whether animals would choose
to be—that is the question. Finally, the passage moves far away from
Johnson, his time, and his island to what was written on the continent
in the preceding century by a French lady. One may properly ask
whether Boswell has in constructing his paragraph contributed "to the
zest of the conversation" or whether he has drifted away from conversation altogether and, like an unscrupulous performer, simply upstaged
the great star. Is Boswell's dramatic method sometimes that of the ham
actor?

Not in this case, certainly, for despite our initial doubts, it is clear
that everything Boswell has done here conspires to produce the illusion
—*effect* is a better term—of a lively, interesting, four-way dialogue
between Johnson, Hutcheson, Boswell, and Madame de Sévigné. That
the dialogue not only never took place, but that it never could have
since two of the "speakers" were dead in 1776, only reminds us that
Boswell's imagination was not turned off by his determination to remain faithful to the truth, to *invent* nothing. There are other effects,
too: finding him in agreement with the praiseworthy author of *Moral
Philosophy* should raise or maintain our esteem for Johnson. Boswell,
by his willingness to praise the moral and intellectual qualities of
Hutcheson even while going on to indicate a deficiency in his (and
Johnson's) statement of the question, has minimized the moral distance
between the narrator, Hutcheson and Johnson: all are worthy men who
can respect one another without falling into dull identity of viewpoint
on an issue. By the same token, moral distance between Hutcheson,
Johnson, and the reader is minimized. Identifying with the biographer

in the absence of any reason here for feeling unlike him, the reader will adopt the narrator's moral kinship with men who are explicitly singled out for praise or implicitly praised by association. Madame de Sévigné, too, is made to seem morally close to all concerned: Boswell carefully characterizes her as a person who "felt with delicate sensibility the prevalence of misery." Along one axis, therefore, aesthetic distance has been sharply reduced.

Along another axis, however, distance is simultaneously increased. As the passage moves from Johnson's sentences to the viewpoints of Hutcheson, Boswell and Madame de Sévigné, the reader is taken further away intellectually from Johnson. His statement of the question is said to be inadequate, the topic is broadened from the misery of animals to the misery of people, and the lady is allowed to have the last word. There is no crushing retort from Johnson—"Madame (frowning)"—to bring readers back under the sway of his position and settle the matter. Nor does Boswell settle it. We are left only with the implication created by his restatement of the question, i.e., that Madame de Sévigné is more nearly right than Johnson.

But it is *we* who must finally decide. Boswell has in effect collapsed the distinction between actor and audience, between action on-stage and action off-stage. His drama—here as elsewhere throughout the *Life* primarily a play of ideas—becomes supremely interesting because he has put into it the most interesting of all possible characters: ourselves. It is a strikingly "modern" piece of dramaturgy. But as Professor Pottle has acutely pointed out, the current popularity of Boswell's journals is no accident due simply to their spicy night-scenes: "Boswell writes like one of us. His style raises few feelings of strangeness in the minds of readers whose taste has been fixed by Maugham, Hemingway, Joyce, Faulkner, Salinger." [7] We are at home with Boswell's style for many reasons, but partly because he can so adroitly manipulate different aspects of aesthetic distance, as in the passage under discussion, to implicate us in his drama by keeping us morally (or emotionally) close to his cast of characters while nevertheless compelling us to stand back intellectually and pass judgement on the argument. Such manipulation does not occur in every scene of the *Life* any more than eloquent soliloquies occur in every act of Shakespeare's plays, but the occurrence is sufficiently frequent to warrant notice as a striking felicity of style. Of course one could read the Ten Commandments and then disagree with them. Any reader is always free to dispute any point. But some works do not *encourage* dissent as Boswell does in passages similar to the one I am discussing. His very deftness in sustaining interest by involving readers in the Johnsonian dialectic accounts for the dearth of critical

[7] Pottle, "Boswell Revalued," 91.

comment on this aspect of his style. His art elegantly conceals itself, for it is only rarely that he makes his invitation as crudely explicit as for example he does when after describing one heated argument he says: "My readers will decide upon this dispute." (III, 350)

Even that comparatively crude invitation, however, serves to make the reader move away intellectually from Johnson, who otherwise would have had the last word in that argument when he silenced Boswell by growling "Nay, if you are to bring in gabble, I'll talk no more. I will not, upon my honour." (III, 350) In many scenes Boswell relies on another device for implicating readers and simulating conversation at that point in the narration where it is made clear that everyone has been reduced to silence by Johnson, all real conversation thereby ceasing. Consider, for example, the evening in 1775 at Cambridge's villa when Johnson, after giving his views on the harmlessness of *The Beggar's Opera*, brought the discussion to an abrupt halt by "collecting himself, as it were, to give a heavy stroke," and saying "There is in it such a *labefactation* of all principles, as may be injurious to morality." (II, 367) Johnson's remark is followed by two paragraphs, the second giving information on the stage history of *The Beggar's Opera* and the first providing the following information:

> While he pronounced this response, we sat in a comical sort of restraint, smothering a laugh, which we were afraid might burst out. In his Life of Gay, he has been still more decisive as to the inefficiency of 'The Beggar's Opera' in corrupting society. But I have ever thought somewhat differently; for, indeed, not only are the gaiety and heroism of a highwayman very captivating to a youthful imagination, but the arguments for adventurous depredation are so plausible, the allusions so lively, and the contrasts with the ordinary and more painful modes of acquiring property are so artfully displayed, that it requires a cool and strong judgement to resist so imposing an aggregate: yet, I own, I should be very sorry to have 'The Beggar's Opera' suppressed; for there is in it so much of real London life, so much brilliant wit, and such a variety of airs, which, from early association of ideas, engage, soothe, and enliven the mind, that no performance which the theatre exhibits, delights me more. (II, 367)

Here only the first two sentences are obviously relevant inasmuch as they finish describing the scene and then relate Johnson's conversation to his writing. Moreover, the first sentence increases our emotional distance from Johnson by showing that even the other actors in the scene found his remark funny. As the butt of ridicule, even silent ridicule, he is moved away from us.[8] This comic distancing also reminds us of

---

[8] *Held up to ridicule* is the stock phrase, its dead metaphor sufficiently suggesting the shift away from us of any comic object.

Johnson's uniqueness, for who but he could ever silence intelligent men by referring to labefactation?

The rest of the paragraph moves us away from Johnson intellectually as Boswell now occupies the stage alone, again soliloquizing: "I have ever thought somewhat differently. . . ." Though the effect is of discussion continued through more pros and cons (since Boswell proceeds to tell us what he has always thought on both sides of the issue), in fact the description of the scene has ended. We are not even given what Boswell thought *at the time* but was perhaps too intimidated to speak aloud; instead we merely have his life-long ambivalent response to *The Beggar's Opera*. The question is, or is intended to be, complicated by Boswell's ruminations, and the reader is thereby presented with a dialectic whereas in fact during the scene described—that evening's conversation at Cambridge's villa—there was only a comical *ipse dixit*.

Boswell has deftly added to the comic interlude an intellectual pleasure. After laughing, the reader must think about whose argument is most convincing. Very often more serious moments are also protracted in the same manner to make readers disengage themselves from Johnson's dicta and assess them. Having reported a conversation during which Johnson gave his views on marital infidelity, for example, Boswell adds a paragraph of disagreement beginning "Here it may be questioned, whether Johnson was entirely in the right." (III, 406) It is we who are left to settle the question. Again, after reporting Johnson's dismissal of *Elfrida* with the concession that it contains "now and then some good imitations of Milton's bad manner," Boswell registers dissent in a paragraph beginning "I often wondered at his low estimation of the writings of Gray and Mason." (II, 335) Having reported Johnson's refusal to concede that the "question concerning the legality of general warrants" was important, Boswell attributes the refusal to Johnson's characteristic "laxity of talking" and then adds that "surely, while the power of granting general warrants was supposed to be legal . . . we did not possess that security of freedom, congenial to our happy constitution, and which, by the intrepid exertions of Mr. Wilkes, has been happily established." (II, 73) By casting his opposition to Johnson in the form of praise for Wilkes, Boswell wrenches us intellectual miles if not light-years away from Johnson. We are of course always free to return. But simply by adding a sentence, Boswell has insured that agreement with Johnson on this issue will not be easy or thoughtless. Any siding with Johnson here that is not mere bias will only occur after the reader has mentally thrashed through the complicated question of Wilkes and liberty.

The list of similar examples could easily be lengthened. More significant than their mere presence as devices for engaging readers as "participants" in the Johnsonian intellectual drama, however, is the

high degree of success Boswell has achieved. It has always been difficult
for critics to remain indifferent to his Johnson. It is Boswell's skill as
much as Johnson's personality that has created so many partisans and
so many detractors. Even those who in Macaulay's vein disparage Boswell are in their way testifying to his effectiveness in forcing commitment, because it has usually been impossible merely to register dislike
of the biographer without also inclining to preference for his subject.
Even in a recent, sympathetic, and utterly unMacaulaian account of
"the self-portrait of James Boswell which emerges from the conversations, letters, and editorial comments of the *Life of Johnson*," Irma
S. Lustig was moved to deplore Boswell's "arrogant posthumous refu-
tations of Johnson's views" on slavery.[9] The corollary of her reaction
is increased respect for the victim of Boswell's arrogance. And whatever in this fashion sustains or creates admiration for Johnson works
towards an important goal of the *Life*. Boswell has created a rhetorical
dilemma from which it is hard if not impossible to escape: agree with
him and your opinion of Johnson, always finally admired by Boswell,
goes up; disagree with or dislike him, and Johnson, by contrast, looks
good.

Without so many Boswellian intrusions after the fact, the dilemma
could not be posed in such acute form. Nor could it always function
so effectively without Boswell's adroit blurring of the distinctions between past and present and between thought and word. In the above
examples it has mostly been clear that Boswell is dissenting from Johnson at a safe distance in time: narrator and reader move away from the
reported scene to its recollection in tranquillity. "*Here* it may be questioned whether Johnson was entirely in the right." Here in the book
and now that he is gone. But not then and there. Often, however, the
line between past and present is not so sharply drawn. After quoting
Johnson's opinion of Rousseau, for example, Boswell has the last word
by adding: "This violence seemed very strange to me, who had read
many of Rousseau's animated writings with great pleasure, and even
edification; had been much pleased with his society, and was just come
from the Continent, where he was very generally admired. Nor can I
yet allow that he deserves the very severe censure which Johnson pro-
nounced upon him." (II, 12) In this case Boswell carefully distinguishes
between his present opinion as he writes the biography and what he
thought when he heard Johnson censure Rousseau. Yet the effect of
so closely juxtaposing two consistently dissenting Boswellian opinions
is to collapse the temporal distance between then and now. What
seemed strange at the time still does. Nothing has shaken Boswell's admiration of Rousseau, which therefore gains at least some weight in

    [9] Irma S. Lustig, "Boswell on Politics in *The Life of Johnson*," *PMLA*, LXXX (Sep-
tember 1965), 393.

our mental scales as it is balanced against Johnson's view. By the same juxtaposition, written word (what Boswell cannot yet allow as he writes the biography) coalesces with thought (what Boswell thought then about the strangeness of Johnson's violence). Similarly, Boswell's reported thought has for readers almost the same effect as disagreement spoken aloud. We see two sides of a "dialogue" whereas a witness of the scene itself (or a tape recording) would have noted only Johnson's remark and Boswell's silence.

Elsewhere Boswell more thoroughly collapses the distance between past and present. After quoting verbatim, for example, Johnson's remarks on Churchill's poetry—remarks incited, Boswell vaguely reports, by his having "ventured to hint that [Johnson] was not quite a fair judge"—the biographer adds: "In this depreciation of Churchill's poetry I could not agree with him. It is very true that the greatest part of it is upon the topicks of the day. . . . But Churchill had extraordinary vigour. . . . Let me add, that there are in his works many passages which. . . ." (I, 419–20) The paragraph from which these extracts are taken moves smoothly from past ("I could not agree") to present tense ("it is very true. . . . Let me add") via a listing of the attributes of Churchill's poetry. Most often any such list will be in the present tense, whether or not it is part of a reported thought or statement that occurred in the past. With such a passage serving as bridge, readers are less aware of the switch in tense. Moreover there is some ambiguity about Boswell's statement that he "could not agree" with Johnson. Does this mean that Boswell was silent? Or does it mean that he spoke aloud arguments like—but not verbally identical with—those that he gives in the paragraph?

These questions are significant precisely because they do not occur to most readers of the *Life*. So deftly has Boswell collapsed the distance between past and present in this and similar passages that we are normally aware only of the effect he thereby creates: dialogue is imitated with elegant artifice. Here, as in other ways elsewhere, Boswell's art achieves its success not by transcription of life but rather by skilful mimesis.

III

Most often temporal distance is minimized without conspicuously lessening intellectual distance between readers and Johnson. Especially is this the case in the many extreme instances where Boswell simply takes his readers completely back into the past by omitting any explicit reference to the act or moment of writing, giving instead merely what Johnson said together with what Boswell was moved to *think* in response. A few examples will suffice: "Here he seemed to me to be strangely deficient in taste; for surely statuary is a noble art of imitation. . . ." (II, 439); "My illustrious friend, I thought, did not sufficiently admire Shenstone. . . ." (II, 452); "I, however, could not help thinking that a man's humour is often uncontroulable by his will" (III, 335); "Seeing him heated, I would not argue any farther; but I was confident that I was in the right. I would, in due time, be a Nestor, an elder of the people; and there *should* be some difference between the conversation of twenty-eight and sixty-eight." (III, 337) In noting these and similar passages where Boswell's dissent maintains significant intellectual distance between readers and Johnson, however, it must be remembered that such distance is always a comparatively short remove from Johnson and from the narrator's reiterated position of affinity to his "illustrious friend." It is somewhat surprising to discover how often Boswell actually disagrees with Johnson, because what the narrator causes to stand out most prominently in our memories of the *Life* are the places where he describes his response to Johnson in such phrases as "I thought I could defend him at the point of my sword. My reverence and affection for him were in full glow." (III, 198)

Historically, most readers have been left with an overwhelming impression of Boswell standing close to Johnson in rapt attention and enthusiastic accord. Boswell's kinship with the master has so far exceeded that of all but the most sympathetic readers as to become proverbial: only recently has *Webster's New International Dictionary* stopped defining *Boswellian* as "Relating to, or characteristic of Dr. Johnson's biographer, James Boswell, whose hero worship made a faithful but often uncritical record of details." [10] But this impression, like the reader's memory of almost unceasing dialogue, is partly dramatic illusion. Boswell has in fact chosen to remain in the middle distance between objective spectator and hero-worshipper. He portrays himself as sufficiently close to the average reader so that the narrator functions as Everyman reacting to the unique Johnson while nevertheless remaining close enough to the Sage in outlook and disposition so that readers will accept the biographer as a fit guide.

Upon the success of Boswell's difficult balancing act depends much of the effectiveness of "the peculiar plan of his biographical undertaking." Imagine for a moment a *Life of Johnson* written in the grand manner of, say, Winston Churchill, Gibbon, or even of Johnson himself. Imagine that all the Johnsonian dialogue is retained together with

---

[10] Even *Webster's Third New International Dictionary* (Springfield 1961) defines *Boswell* as "One who out of admiration or hero worship records in detail . . . the life . . . of a famous or otherwise significant contemporary. . . . One who stays in almost constant attendance upon another out of great admiration or hero worship, often in a voluntarily servile position."

the distinctive piling up of "each scene" of which there is some record. But suppose that for Boswell's personality *as it is revealed to us in his book* the personality of Gibbon, or Churchill, or some twin of Johnson were substituted. There is no reason why such a *Life* could not be successful. But it would be a radically different kind of success. We as readers would be kept at a greater distance from the subject by the imposing personality of the narrator. There would always be the distancing realization, as there is for example in reading Johnson's lives of the greater poets, that we are watching from a respectful distance one extraordinary mind respond to another. The spectacle may be fascinating, but we cannot become part of it in the way Boswell makes us participants in the drama of Johnson's life, living "o'er each scene *with him.*"

To maintain the narrator equidistant from readers and Johnson, Boswell plays many roles throughout the *Life.* Extremes are intended to cancel each other out. Sometimes he is simply, like his presumed readers, the literate Everyman intelligently in touch with the current state of *belles lettres.* Thus, early in the *Life,* commenting on the publication of *London,* he remarks: "To us who have long known the manly force, bold spirit, and masterly versification of this poem, it is a matter of curiosity to observe the diffidence with which its author brought it forward into publick notice." (I, 123–24) Here by his choice of plural pronoun Boswell joins the group of readers long familiar with one of the century's most famous poems. Distance between narrator and audience is collapsed to the point of complete identification. He becomes one of us. Much later in the *Life* Boswell quotes a passage of self-analysis written in his journal *after* an evening with Johnson, asserting in defense of the quotation: "This reflection, which I thus freely communicate, will be valued by the thinking part of my readers, who may have themselves experienced a similar state of mind." (III, 225) Together with its attempt to disarm through bullying and flattery any charge of irrelevancy, this sentence endeavors to move narrator and reader close together intellectually by suggesting a probable kinship of mental experience. Boswell thus reassures his intelligent readers that he is like them.

He usually does so less explicitly by disclosing the narrator's character in ways that invite readers to infer their likeness to him. After quoting, for example, Johnson's statement that the most famous men worry most about losing their reputation, Boswell adds: "I silently asked myself, 'Is it possible that the great SAMUEL JOHNSON really entertains any such apprehension, and is not confident that his exalted fame is established upon a foundation never to be shaken?'" (I, 451) Here the narrator portrays himself as an ordinary, unfamous man startled to catch a close glimpse of the uncertainties of greatness. Readers not suffering delusions of grandeur will by this sentence and similar ones be moved closer to the narrator and thereby encouraged to continue their identification with him, seeing Johnson through eyes that might —so Boswell makes us feel—belong to any man.

To ensure, however, that such identification by readers with the narrator does not become so close that confidence in him as fit guide is rattled by any feeling that he is *just* an ordinary man, no different from the run-of-the-mill reader of biographies, Boswell frequently invites attention to his own distinctiveness. At no point in the *Life* are we allowed to forget that the narrator, unlike his readers, was after all one of the charmed circle admitted to friendship with "the great SAMUEL JOHNSON." . . .

Not that all members of Johnson's circle are to be equally respected. Boswell's reiterated sniping at Hawkins and Mrs. Thrale is only part of a sustained effort intended to display the narrator as remarkable both for his illustrious friendships and for his unusually selfless devotion to the exacting task of accurately portraying Johnson. After briefly marvelling at his own talents, Boswell says in the Advertisement to the first edition that he "will only observe, as a specimen of my trouble, that I have sometimes been obliged to run half over London, in order to fix a date correctly; which, when I had accomplished, I well knew would obtain me no praise, though a failure would have been to my discredit." (I, 7) Elsewhere in the *Life* Boswell's painstaking humility is brought to our attention. No seeker after praise, the narrator is only "desirous that my work should be, as much as is consistent with the strictest truth, an antidote to the false and injurious notions of his character, which have been given by others." (III, 391) Though "others" have been slanderous, Boswell is simply the *vir bonus* motivated in all matters by "my earnest love of truth." (III, 147) But his humility does not of course exclude selfless literary courage in a worthy cause: "To please the true, candid, warm admirers of Johnson, and in any degree increase the splendour of his reputation, I bid defiance to the shafts of ridicule, or even of malignity." (III, 190) Brave Boswell! It is not everyone— certainly not every reader, however well disposed to Johnson—who would feel equally willing to expose himself for the cause. As in these and other similar statements Boswell displays his moral and intellectual fitness for his biographical task he is also precluding the possibility of excessive identification by readers with the narrator. Though similar, they are not to be thought identical. He has written the book. They have not. At the outset Boswell calls attention to this elementary though crucial distinction by asserting: "The labour and anxious attention with which I have collected and arranged the materials of which these volumes are composed, will hardly be conceived by those who read them with careless facility." (I, 5–6) Readers speeding through the *Life* are thus reminded of their distance from its author. So far are they from him intellectually that his immense labors will be almost beyond their ken.

Another way of avoiding over-familiarity with readers is to put on the mask of Johnsonian sage, however ill-fitting. Boswell moves away from us—in the direction of Johnson—by assuming that we need advice and then supplying it. "The excellent Mr. Nelson's 'Festivals and Fasts' . . . is a most valuable help to devotion; and in addition to it I would recommend two sermons on the same subject, by Mr. Pott, Archdeacon of St. Alban's, equally distinguished for piety and elegance." (II, 458) Here readers are reduced to students taking their reading list from the pious, learned narrator.

Boswell also affects the Johnsonian tone by sententiously generalizing on the human condition: "To such unhappy chances are human friendships liable. . . ." (III, 337) Or, for another example, after stating his disagreement with the philosophy of *Rasselas* and then affirming that life is sometimes more, and sometimes less happy, Boswell launches another flight of sober religious advice by saying: "This I have learnt from a pretty hard course of experience, and would, from sincere benevolence, impress upon all who honour this book with a perusal, that until a steady conviction is obtained that the present life is an imperfect state, and only a passage to a better. . . ." (I, 343–44) Nothing better illustrates the mask Boswell is trying to wear in such passages than his soliloquy on "how different a place London is to different people." Placing himself momentarily in their shoes, the narrator concisely describes how the city will seem to "a politician . . . a grazier . . . a mercantile man . . . a dramatick enthusiast . . . a man of pleasure" and, finally, to such men as the narrator himself: "But the intellectual man is struck with it, as comprehending the whole of human life in all its variety, the contemplation of which is inexhaustible." (I, 422) Throughout the *Life* Boswell tries to portray himself as intellectual man, just as throughout the *London Journal* he tries equally hard to cast himself in a very different role: "The description is faint; but I surely may be styled a Man of Pleasure. . . . I patrolled up and down Fleet Street, thinking on London, the seat of Parliament and the seat of pleasure, and seeming to myself as one of the wits in King Charles the Second's time." [11]

Equally illustrative of the role Boswell adopts in his *London Journal* —and of the fact that he does conspicuously put on different masks to match different rhetorical situations in different books—is the moment

[11] James Boswell, *Boswell's London Journal*, ed. Frederick A. Pottle (New York 1950), 140.

when he reports that he "drank about and sung *Youth's the Season* and thought myself Captain Macheath." [12] In the *Life* our narrator does not try to persuade readers that he is a Restoration rake or a character in search of a comic opera. Instead there are the solemn warnings and sober generalizations together with a very different and much more respectable order of literary allusions. Thus after describing how he has educated his sons, Boswell adds his expectation that "they will, like Horace, be grateful to their father for giving them so valuable an education." (III, 12) We readers may well feel put back in our (distant) place by such glimpses of the generalizing Scottish sage as intellectual man in London or at home raising so wisely his Horatian family.

Or we may smile. And in that case Boswell's control of aesthetic distance has wavered. Instead of moving away from readers in the direction of Johnson, the narrator has moved away in the opposite direction, far alike from Johnson and from the readers to whom the narrator has become an object of ridicule. Consider, for example—everyone will have his own favorite—Boswell at grips with the problem of getting up in the morning: "I have thought of a pulley to raise me gradually; but that would give me pain, as it would counteract my internal inclination." (III, 168) Here the would-be intellectual man looks suspiciously like Falstaff. Too often throughout the *Life* Boswell comes onstage wearing the wrong mask.

Control of aesthetic distance has not wavered every time we smile at the narrator, however. After one memorable Johnsonian retort, for example, Boswell interrupts his narrative by observing: "I never heard the word *blockhead* applied to a woman before, though I do not see why it should not, when there is evident occasion for it." (II, 456) Here readers may well laugh at Boswell's naïvely solemn consideration of the great lexicographer's word usage. But the narrator has conspicuously placed tongue in cheek to heighten our appreciation of a comic incident. While pausing to smile at the narrator readers must also stand apart from what Johnson has just said and reflect on how extraordinary the remark really was. It is only when our laughter at Boswell deflects attention from Johnson that control of aesthetic distance has been lost. All readers would not agree on precisely where and how often these mistakes occur. Nor is it necessary to achieve such accord. The acceptance for so long of Macaulay's response to Boswell is sufficient evidence of the *Life's* most conspicuous flaw: Boswell's characterization of the narrator does not always keep readers at a respectful distance which is nevertheless sufficiently close to him for that identification which induces maximum involvement in the *Life's* play of ideas. Where the *Life* fails it is because we are allowed to come too close to the narrator or,

[12] *Ibid.*, 264.

what is in effect perhaps the same thing, because he is pushed too far away from us in the wrong direction.

But to pinpoint the *Life's* weak spot as a wavering in control of aesthetic distance between readers and narrator is not to say that the weakness is fatal. Boswell's overwhelming achievement in creating the most famous biography in the English language is in large though not exclusive measure due to his skill in varying different aspects of aesthetic distance.[18] The places where his control falters are comparatively few and, because he does succeed in establishing the rhetorical dilemma which I have described above, even those places do not significantly undercut the *Life's* over-all effect. The brief examples discussed here typify the ways in which by varying aesthetic distance along several axes Boswell succeeds in sustaining interest, maintaining faith in the narrator, creating sympathy for Johnson, and, perhaps most important of all, preventing the reader's sensibilities from being anesthetized by such thorough immersion in the "Johnsonian aether" as Boswell's plan entails. By compelling us so often to stand back and weigh Johnson's distinctive remarks, by thus reminding us of how debatable his views so often are, Boswell keeps alive our sense of wonder. We are never allowed to forget that Samuel Johnson was "a man whose talents, acquirements, and virtues were so *extraordinary,* that the more his character is considered the more he will be regarded by the present age, and by posterity, with admiration and reverence."

[18] I have called Boswell's work the most famous rather than simply the greatest biography in our language in order more thoroughly to beg the questions raised by Donald J. Greene, "Reflections on a Literary Anniversary," *Queen's Quarterly,* LXX (1963), 198–208. [Included in this volume, p. 97—ED.] While Greene's invitation to ask ourselves whether the *Life* is a biography at all deserves a careful response, it has not seemed necessary to provide one here. The question of the book's adequacy as a record of Johnson's life is distinct from—though of course not unrelated to—the question of how Boswell disposed his material in ways that enabled the *Life* to achieve its enduring and undeniable success.

The *Life of Johnson*: Art and Authenticity

by Frederick A. Pottle

"What I consider as the peculiar value of the following work," wrote Boswell in the paragraphs introductory to the *Life of Johnson,* "is the quantity that it contains of Johnson's conversation." The emphasis is on the word "peculiar," and in what follows I shall retain that emphasis. Boswell's strategies as a biographer, his handling of conversation apart, are impressive and worthy of analysis, but they are not strikingly different from those employed by other biographers. His conversations, by general consent, are unique, and they do constitute the supreme value of his work.

Study of Boswell's journal in the forty years or so since its recovery has added nothing to the value of the *Life* as a work of art, but it has revolutionized our thinking about the way in which the book was written and the nature of Boswell's genius. Gone forever is the assumption that he subordinated the claims of his own life to the recording of Johnson's; gone is the picture of him trotting at Johnson's elbow with a notebook, anxiously jotting down Johnson's conversation for use in a biography. We know now that he was an inordinately ambitious man who lived his own life fully; we know that with only trifling exceptions he recorded the conversations in his journal, sometimes long after they occurred. And we know that he put the conversations in the journal primarily because they were a most important part of his own life.

Boswell was remarkable for his zest in life and his consciousness of enjoyment. The first of these traits is common, the second rare. And he added to these a third trait so uncommon as to appear unique. He did not feel that his zest had been exhausted, his enjoyment fully enjoyed, till he had made a lively *written* record of it, till he had journalized it. "I should live no more than I can record," he wrote, "as one should not have more corn growing than one can get in." [1] He was in London when he made that observation, but the context makes clear

---

"The Life of Johnson: Art and Authenticity*" by Frederick A. Pottle. A previously unpublished essay written specially for this collection, Autumn 1969. Printed by permission of the author.*

[1] Journal, 17 March 1776, *Boswell: The Ominous Years, 1774–1776,* ed. Charles Ryskamp and F. A. Pottle (New York: McGraw-Hill, 1963), p. 265.

that Johnson's conversation was only the contribution of his best field to the bumper crop he was having difficulty in getting under cover. As Geoffrey Scott has said, the Johnsonian portions of Boswell's journal are not different in kind from the rest; they flow in and out with no change whatever in method or emphasis. "The vast, bracing difference is the subject matter." [2]

Boswell's compulsion to record his own life is to my knowledge unique for urgency and sharpness of definition, but one does not have to look far for an illuminating parallel. If a man wrote, "I should construct no more plots than I can write down," we should know at once that we had to do with a practicing author, and probably with a successful one. Well, Boswell felt in the day-by-day happenings of his own mind the overriding significance which novelists feel in their inchoate fictions; and he felt the same pressure that they do to bring *his* matter to full literary expression. *Literary* (or, if you prefer, *imaginative*) expression is important, for we must not suppose that Boswell's recurring stretches of terse, dry chronicle or of abbreviated and cryptic notes were felt by him to be anything more than ground-holding operations. Gone with the obsequious Boswell and the notebook Boswell is the amateur Boswell, the Boswell who embarked on serious writing at the age of forty-five. The journal from its earliest period shows the alert awareness of literary problems that marks a gifted and practiced writer. "I observe continually," he wrote at the age of twenty-nine, "how imperfectly upon most occasions words preserve our ideas. . . . In description we omit insensibly many little touches that give life to objects. With how small a speck does a painter give life to an eye!" [3] Boswell not only had literary imagination to a high degree, but his imagination worked specifically and continually in literary recording of the daily matter of his own experience, including Johnson. It made an incalculable difference to the quality of the *Life* that he did not merely store notes and wait till he was middle-aged before he attempted the full imaginative expression of the conversations which were to shape and to dominate his book. Thanks to his journalizing, he had begun applying his best literary powers to Johnson at the age of twenty-two, with twenty-two years left in which to test and improve his method.

"Little touches that give life to objects"—one finds no language of this kind in the "Advertisements" to the *Life* or in the introductory paragraphs in which Boswell discusses his method. He firmly claims

[2] Geoffrey Scott, *The Making of the Life of Johnson*, vol. 6 of *Private Papers of James Boswell from Malahide Castle in the Collection of Lt. Colonel Ralph Heyward Isham*, ed. Geoffrey Scott and F. A. Pottle (Mt. Vernon, N.Y.: privately printed, 1929), p. 67. Scott's brilliant study underlies everything of any value that has been written on Boswell's method.

[3] Journal, 16 September 1769, *Boswell in Search of a Wife, 1766–1769*, ed. Frank Brady and F. A. Pottle (New York: McGraw-Hill, 1956), p. 292.

credit for "labour and anxious attention" in collecting and arranging his materials, for "stretch of mind and prompt assiduity" in preserving conversations, and for the "degree of trouble" he had put himself to to "ascertain with a scrupulous authenticity" the "innumerable detached particulars" of which his book consists. Nothing more. I suppose even a vainer author might hesitate to proclaim his artistry in advance of the verdict of the public, but I suspect another motive for silence. He was determined above all things that readers should not only grant his claim of scrupulous authenticity in the detached particulars, but should also assume an equal degree of authenticity in his overall depiction of Johnson; and he may have feared that any talk about his *giving* life to objects would imply that he had colored fact with fiction. Whatever the motive, he gives, and seems to have intended to give, the impression that all that was needed to produce the *Life* was a remarkable memory and lots of hard work. It is this impression, no less erroneous for having been initiated by Boswell himself, that I now wish to demolish.

Boswell, by birth or self-training or both, did have a remarkable memory. Furnished with the right kind of clue and given time, patience, and freedom from distraction, he could bring back any desired portion of his past with a wealth of authentic detail; particularly, could recover a good part of what he and other people had really said on that occasion. The right kind of clue was a written note made by himself; nothing else would serve. Without such a note, his memory was no better than anyone else's. I think he would have preferred to post his journal soon enough after the events so that no note would have been necessary, but there were few periods in his life when he could journalize with that degree of regularity. His more general practice was to make rough, abbreviated notes on scraps of waste paper soon after the events, the sooner the better. From these he wrote up the journal days, months, even years later, as he could find time.

The notes often seem quite disorderly and unselective, as though Boswell, in a tearing hurry, were jotting down whatever rose first to consciousness. When he had expanded the notes in a journal, he almost invariably threw them away. This is of crucial importance in any discussion of his method. If he never got around to write the journal, he would cite the notes as authority, but he never assigned to notes an authority superior to that of the journal based on them. *Faute de mieux*, the notes were a record ("the bones," he himself said) of his life, but their primary and essential function was to serve as hints for remembering. ("A hint such as this brings to my mind all that passed, though it would be barren to anybody but myself.")[4]

---

[4] Journal, 9 January 1768, *ibid.* p. 115. I do not remember where Boswell called a stretch of notes "the bones of my most wretched existence," but I guarantee the authenticity of my recollection.

The process of recollection did not stop with the journal, but continued to operate when matter from the journal was transferred to the *Life*. The greater part of the extended Johnsonian conversations involving several speakers, it now appears, had never been expanded in the journal at all, and had to be worked up from notes jotted down many years before. One also frequently finds Boswell adding sentences and paragraphs to portions of fully written journal. Some of these additions seem to be authentic but undated recollections for which he had to find plausible points of attachment; others, I have no doubt, are a second crop of memory, gathered as he relived the matter he had copied.

Boswell's insistence on the essentiality of circumstantial detail in all recording of the past was probably in part due to his own need for a body of authentic and unique historical detail if he were to set his memory in successful operation. Circumstances were a bridgehead into the forgotten country; if he held the bridgehead, he could reoccupy the country at will. And reoccupation was an effort, was by no means automatic. It now appears probable that some—perhaps all—human brains do store specific physical traces of massive coherent quantities of past sense-experience, and that appropriate stimulus will cause a person to relive such clusters in minute detail, for example, will enable the stimulated person to listen in on an old conversation in its entirety, or hear an orchestra play through a piece of music just as he heard it years before.[5] Boswell's memory was almost certainly not of that sort. So far as our present-day knowledge goes, that kind of recall requires physical stimulus of an abnormal sort—an adventitious current of electricity. Besides, Boswell's recall, as I hope to demonstrate, does not bear the marks of passive or rote memory. He seems rather to have been a champion in a game we all play at all the time; and his performance appears to be no less explicable in terms of genetic aptitude and sedulous training than the performance of any other champion. Many examples of equally extraordinary memory have been cited. William Lyon Phelps says that his older brother Dryden could remember some definite thing that had happened to him on every day of his life after the age of five. If you asked him, "What did you do on February 17, 1868?" he would ponder about twenty seconds and then say, "That was a Monday," and in about three minutes he would describe the weather on that day and tell you something he had done on it.[6]

---

[5] Wilder Penfield, M.D., "Some Mechanisms of Consciousness discovered during Electrical Stimulation of the Brain," *Proceedings of the National Academy of Sciences* XLIV (15 Feb. 1958), 51–66. Dr. Penfield remarks of his patients' recovery of the past, "This is not a memory, as we usually use the word, although it may have some relation to it. No man can recall by voluntary effort such a wealth of detail."

[6] William Lyon Phelps, *Autobiography with Letters* (New York, Oxford Univ. Press, 1939), pp. 367–368.

The fixing of the day of the week which seemed so remarkable to Phelps was a simple arithmetical calculation which many people could perform in their heads if they knew the formula, but back of the recall of the events of the particular day undoubtedly lay extended special training of some sort. When I ask myself what could have been the groundwork for that training, I get a mental picture of a shelf full of little books—a long unbroken row of calendar diaries, begun at the age of five and kept without a single blank for more than sixty-five years. At any rate, Boswell's journal, with its subsidiary notes and memoranda, was not merely a record of the past; it was also persistent and scrupulous training in recollection.

Boswell never maintained that his records of conversation were complete word-for-word transcripts of what was said on a given occasion, though perhaps by silence he encouraged readers to think that they were. They are obviously epitomes or miniatures: people talking for the length of time he says or implies that they did talk would certainly have uttered many more words than appear in his account. If we had a tape-recording of the originals of any of the conversations in the *Life,* we should find that the progression of the dialogue was by no means as swift and economical as the *Life* represents it to have been. Real speakers in real life—even Johnsons—wander off in side-excursions and bog down in irrelevancies. The conversations of the *Life* are in this respect not unlike the brief reports of long extemporaneous speeches in Parliament that one finds in eighteenth-century newspapers and magazines. The reporter has condensed the speeches in language which he was not given verbatim but had to find for himself, yet he has infused the styles of the speakers into his condensations.

One can get a nice definition of Boswell's method by combining two remarks which he himself made. At the end of the first extended conversation he recorded in his journal—a conversation which he and his friend Andrew Erskine had with David Hume on 4 November 1762— he congratulated himself on having "preserved the heads and many of the words" of a dialogue lasting an hour and a half. His record comprises less than a thousand words.[7] And he says (this time in the *Life of Johnson*) that when, in the course of time his mind became *"strongly impregnated with the Johnsonian aether,"* he could carry Johnson's remarks in his memory and commit them to paper "with much more facility and exactness."[8] This answers a number of questions which have been put as to the historical veridicality of his Johnsonian record.

---

[7] Journal, 4 November 1762, *Private Papers,* as in n. 2 above, i (1928), 126–129. Or pp. 100–104 of the deluxe edition of *Boswell's London Journal, 1762–1763* (London: Heinemann, 1951).

[8] 1 July 1763: James Boswell, *The Life of Samuel Johnson,* ed. G. B. Hill, rev. L. F. Powell (Oxford: Clarendon Press, 1934–1964), I, 421.

Did Boswell ever follow Plato's practice, inventing for Johnson extended dialogues which were appropriate but non-historical? No. Boswell's Johnson is always "authentic": his Johnson speaks on the major topics ("heads") of conversations actually entered into by the historical Johnson. Did Boswell, while sticking to the historical "heads," ever allow himself to extend Johnson's remarks by fictitious matter serving purposes of Boswell's own? No. Boswell's Johnson is in this respect "*scrupulously* authentic." He does not say *all* that the historical Johnson said on a given occasion, but he says nothing that the historical Johnson did not in substance say—on that occasion or another. For Boswell indeed admits in the journal (unfortunately not in the *Life*) that he does not always observe strict chronology in recording Johnson's remarks. These admissions, I believe, all occur in journals like the Hebridean journal of 1773 or the Ashbourne journal of 1777, on occasions when he was continuously in Johnson's company, had fallen behind in his effort to keep the journal up to the preceding day, and feared that he would lose some vivid recollection if he put off recording it till its proper date was reached.[9] To me the fact that the admittedly displaced sayings fit their surroundings so perfectly suggests that an additional principle may have been operating, at least part of the time. In all conversation between intimate friends, the same topics keep recurring; and, when they do recur, most speakers repeat themselves almost automatically and almost verbatim. Johnson's "exuberant variety" [10] generally protected him from flaccid repetition; and Boswell, when the same topic occurred twice in matter still unjournalized, may sometimes have conflated the two versions in the journal, using the portions of each which at the time he considered superior. Or, when he felt sure that he had already reported Johnson on a given topic, he may sometimes not have put the second version into the journal at all. The unrecorded variants, I suggest, hung in his memory as authentic Johnsoniana, and in the *Life* sometimes replaced or were added to the readings of the journal he had before him. This is admittedly speculation, but it is, I believe, the speculation likely to be advanced by anyone who has patiently worked through the vast body of documentation underlying the *Life*.

Does Boswell, then, report Johnson's conversation verbatim? In particular sentences and in some brief passages of an epigrammatic cast, yes. In general, no. The crucial words, the words that impart the peculiar Johnsonian quality, are indeed *ipsissima verba*. Impregnated

[9] *Boswell's Journal of a Tour to the Hebrides with Samuel Johnson, LL.D.*, ed. F. A. Pottle and C. H. Bennett (New York: McGraw-Hill, 1936, 1961), pp. 140, 151, 272, 329; Journal, 15, 17 September 1777, *Boswell in Extremes, 1776–1778*, ed. C. Mc. Weis and F. A. Pottle (New York: McGraw-Hill, 1970), pp. 151, 156.

[10] From Boswell's sentence about the "Johnsonian aether," above, n. 8.

with the Johnsonian ether, Boswell was able confidently to recall a considerable body of characteristic diction. Words entail sense; and when elements of the remembered diction were in balance or antithesis, recollection of words and sense would almost automatically give "authentic" sentence structure. But in the main Boswell counted on impregnation with the Johnsonian ether (that is, on an understanding, grown intuitive, of Johnson's habits of composition) to help him consciously to construct epitomizing sentences in which the *ipsissima verba* would be at home. His greatest virtue as an imitator or re-creator of Johnson's style was not to overdo the idiosyncrasies. Long stretches of his journal, recorded as indirect discourse, were converted into Johnsonian utterance with no more revision than a change of pronouns and tenses.

"With how small a speck does a painter give life to an eye!"—we are back to Boswell's dictum, better able, I hope, to consider its import after this excursus on his memory. Johnson's conversation as Boswell reported it, is, for all its veridicality, an imaginative reconstruction, a re-creation; it is embedded in a narrative made continuously lively by unobtrusive specks of imagination. The stylistic unity of the *Life* has not been enough remarked on. The conversations in the *Life* melt into the narrative; one light of imagination pervades the whole. The Johnsonian ether, which I have just defined narrowly as "an understanding, grown intuitive, of Johnson's habits of composition," was, at a deeper level, a massively detailed conception of Johnson's character, operating to shape into unity all the multifarious and potentially discordant elements of a very long book.[11]

A biographer who aims at this kind of unity (that is, who aims at literature) must win and keep control of his book. The view of the subject's character presented must be *his* view, not the subject's. There is no doubt that the eye that has seen Johnson in the *Life* is Boswell's, but at first glance it is a little hard to see how he managed to keep it so. In incorporating so many complete and largely unselected letters of Johnson, for example, he took a great chance, the chance that so much of Johnson's own strong and idiosyncratic style, so much of Johnson's own casual comment on casual happenings, would force open the focusing tension and reduce the book to a mere compilation. But Boswell counted on the conversations to dominate and control the letters, and he did not trust in vain. In the conversations, as we have seen, he remembered the heads and the very words of a great part of what John-

---

[11] See Ralph W. Rader, "Literary Form in Factual Narrative: The Example of Boswell's *Johnson*," in P. B. Daghlian, ed., *Essays in Eighteenth-Century Biography* (Bloomington: Indiana Univ. Press, 1968), pp. 7–9 and *passim*. Most of the ideas of the present paper were developed in a correspondence with Professor Rader about his pioneering study.

son had actually said on many occasions. But the heads had more often than not been proposed by himself and were of intense personal concern to himself, and the whole had been sifted by *his* memory and vitalized by *his* imagination. He had reconstructed the conversations in the first place to complete portions of his own life in which he had felt himself to be living most fully, and to savor that completeness. The conversations, though they appear to be pure Johnson, are in fact the quintessence of Boswell's view of Johnson.

PART TWO

General Commentary—Strengths and Weaknesses

The Boswell Formula, 1791

by Sir Harold Nicolson

The Boswell Legend

I ended my last lecture with the word "actuality." It is with the same word that I should wish to begin my study of the Boswell formula. For James Boswell invented actuality; he discovered and perfected a biographical formula in which the narrative could be fused with the pictorial, in which the pictorial in its turn could be rendered in a series of photographs so vividly, and above all so rapidly, projected as to convey an impression of continuity, of progression—in a word, of life. Previous biographers had composed a studio portrait, or at best a succession of lantern-slides. Boswell's method was that of the cinematograph. In perfecting his experiment he showed singular originality and surprising courage. He well deserves the central position which he and his formula must always occupy. But the problem of Boswell cannot be elucidated solely by the appreciative method. We must dissect and isolate; we must begin by isolating Boswell from his own legend; then, and then only, will it be possible to define what exactly was his contribution to the art of English biography. . . .

The construction of the *Life of Johnson* may, at first sight, appear artless; yet great art was required to fuse into some coherent and readable whole the disordered mass of notes and letters which Boswell had accumulated. The *Journal of a Tour to the Hebrides* fell almost naturally into shape, since its outlines and internal divisions were dictated by the duration and stages of the journey itself. In composing the *Life,* however, Boswell was from the outset faced with the problem whether he should write a formal biography like that of Hawkins, or mere *Johnsoniana* like the anecdotes of Mrs Thrale. He decided to combine

---

"*The Boswell Formula, 1791*" by Harold Nicolson. From chapter IV of The Development of English Biography (*London: The Hogarth Press, 1927*), *pp. 87–108. Reprinted by permission of The Hogarth Press and Mr. Nigel Nicolson.*

the advantages of both methods. The fact that this decision did not utterly destroy the unity of his book proves that Boswell possessed a very remarkable talent for construction and great literary tact. Consider the technical difficulties. Boswell set out to paint on the large canvas of a full-length biography the "Flemish picture" which he desired to compose. It must be remembered that of the seventy-five years of Johnson's life Boswell had direct knowledge of only twenty-one, and that during these twenty-one years he was only in Johnson's company on two hundred and seventy-six days. He thus possessed but shadowy and indirect knowledge of two-thirds of Johnson's life, whereas his material for the remaining third was, although only in patches, embarrassingly detailed. He endeavoured to conceal this discrepancy by the introduction of letters and the blurring of dates. The skill with which the indirect method of the earlier portrait is dovetailed into the direct and vivid manner of the later period is indeed remarkable. We scarcely realise, when reading the book, that out of a rough total of 1250 pages, 1000 are devoted to Johnson after he had met Boswell, and only 250 to the pre-Boswell period. The book, moreover, is written without prescribed divisions or chapters, and yet its interest, its unity of impression, its sheer limpid continuity is sustained throughout. For the *Life of Johnson* is a work of art, not merely in its actual excellence of outline, but in the careful adjustment of internal spaces. We have thus the absence of comment, or rather the very skilful interspacing of comment—the way in which Boswell first provides the evidence, and then, at a later period, confirms by comment the conclusion which the reader had already reached. I would refer you, as a particular instance of this method, to his treatment of Johnson's strange gullibility on all supernatural matters, and his obstinate scepticism in all natural matters. Boswell tells without comment a story of Johnson's belief in ghosts; a few pages later he tells, equally without comment, of his scepticism regarding some quite natural novelty such as stenography; it is not till much later that he comments directly on his strange conjunction of scepticism and gullibility; and by then the reader can recollect and recognise the evidence on which this comment is based. Equally skilful is his manipulation of the elements of surprise and recognition, of expectation and satisfaction. He keeps his reader constantly in mind, and as constantly pays subtle compliments to his memory and his intelligence. He throws out something, such as the story of Johnson and the orange-peel, which he slyly knows will excite curiosity; he then drops the subject; and then, slyly, he returns to it several pages later, knowing well that greater pleasure will be caused if curiosity is not immediately allayed. This is something more than mere adroitness; it is constructive talent of the highest order. Consider also his sense of values; the skill with which he records the conversation of other people

to the exact degree necessary to explain and illustrate the remarks of
Dr Johnson; the tact with which, while conveying an intimate picture
of himself, he does not obtrude unnecessarily; and how, in the serious
passages on Johnson's last illness, he withdraws with unexpected del-
icacy from the scene. Consider also his very exquisite handling of
cumulative detail; the mastery with which the portrait of Johnson is
conveyed by an accumulation of slight successive touches until the
whole rolling, snorting, rumbling bulk of the man becomes visible,
and we feel that he has grown in intimacy as the book proceeds; that
we have become aware, quite naturally, of his brown stockings, his disordered buttons, the dust settling in his wig as he bangs two folios together, the way he cut his nails, of his servants, his teapot, and his cat.
And this rapid method of portrayal was certainly deliberate. "It appears to me," he wrote to Bishop Percy, "that mine is the best plan of
biography that can be conceived; for my readers will as near as may be
accompany Johnson in his progress, and, as it were, see each scene as it
happened." It is indeed amazing that Boswell should have succeeded
so triumphantly. He was, during the whole period when he was writing
the book, distracted by ill-health, by prolonged dissipation, and by
acute financial and domestic troubles. It is true that he was assisted by
Malone, but the latter was engaged at the time with his own edition of
Shakespeare, and can in no sense be considered as more than a discern-
ing proof-reader. The credit of Boswell's *Johnson* belongs to Boswell
alone. His work was a deliberate and highly successful innovation in
the art of biography. In what exactly did this innovation consist?

Boswell's Originality

The several elements which compose Boswell's method had all been
attempted before. It was Johnson himself who had invented and perfected the method of truthful portraiture and of the realistic biography.
The device of introducing original letters and documents was as old as
Eadmer, and had been exploited by Mason. The device of introducing
anecdote and actual conversations had been brought to a high pitch of
perfection in the French *ana,* had been employed in the *Table Talk of
Selden,* and had been admirably applied to Pope and his circle by
Spence. Boswell's originality was not that he invented any of these
mechanical aids to biography, but that he combined them in a single
whole. That, at least, had never been done before. Nor was this his
innovation due to any accident; it was perfectly self-conscious and deliberate. What he calls "the peculiar plan of this biographical undertaking" had remained in his mind for over twenty-five years. He experi-
mented with it, not very successfully, in his early Corsican journal; he

gave it a trial in his *Journal of a Tour to the Hebrides,* which he published in 1785. Much of the latter had been read by Johnson himself,
and Boswell had profited by his criticism, as he profited by the subsequent criticisms of the public. The notes which he accumulated during
the twenty-one years of his acquaintance with Johnson were continually
being sifted and remodelled. He perfected his method. "I found," he
writes, "from experience that to collect my friend's conversation so as
to exhibit it with any degree of its original flavour, it was necessary to
write it down without delay. To record his sayings after some distance
of time was like preserving or pickling long-kept and faded fruits or
other vegetables, which, when in that state, have little or nothing of
their taste when fresh." [1]

But it was not merely that Boswell perfected the annotative and the
analytical methods of biography. His great achievement is that he combined them with the synthetic. He was able, by sheer constructive force,
to project his detached photographs with such continuity and speed
that the effect produced is that of motion and of life. It is this that I
mean by "the Boswell formula"—a formula which, in the present generation, aided by our familiarity with the cinematograph, might well
be still further developed.

His Courage

Boswell's claim to be the greatest of English biographers is thus
justified not merely by the entertainment which his work provides, but
by the fact that it represents a deliberate and extremely difficult com-
bination of methods, that he invented a highly original and fruitful
formula. I would wish before finishing this lecture to do justice to
Boswell's courage in persisting in his own method. For people were
already becoming alarmed at the growing public taste for truth. They
were alarmed by Curll's ventures, they were seriously alarmed by
Spence. Peter Pindar's *Bozzy and Piozzi; or The British Biographers,*
dates from 1786, and in the following year Canning attacked Boswell's
method in the *Microcosm.* Dr. Waldo Dunn, to whose work on English
biography I have been frequently indebted, has unearthed an even
more specific attack which dates from 1788. "Biography," wrote a
certain Mr Vicesimus Knox, "is every day descending from its dignity.
Instead of an instructive recital, it is becoming an instrument to
the mere gratification of an impertinent, not to say malignant,
curiosity. . . . I am apprehensive that the custom of exposing the
nakedness of eminent men of every type will have an unfavourable

[1] Boswell's *Life of Johnson,* Everyman edition, vol. ii, p. 14.

influence on virtue. It may teach men to fear celebrity." These attacks, it must be realised, were delivered at a moment when Boswell, although ill and tried by domestic trouble, was composing his masterpiece. And Boswell persisted.

The Unknown Johnson

by A. S. F. Gow

For nearly a hundred years, or, to be precise, since September 1831, it has been a commonplace among educated Englishmen that Samuel Johnson is better known to us than any man in history. His vigorous conversation, his oddities of habit and of person, every circumstance of his daily life, are, in Macaulay's phrase, as familiar as the objects with which we have been surrounded from childhood. We know Johnson as we know few even of our most familiar friends. But to this intimacy of knowledge there are nevertheless restrictions which it is the purpose of this paper to consider. Some are obvious, others less so: but one at least leaps to the eye and was justly noted by Macaulay. With that I will deal first.

Our intimacy with Johnson is, in the aggregate, due to many informants, but, among the many, three stand apart. If Boswell is alone in the first class, Mrs. Thrale and Fanny Burney are equally alone in the second. Compared with these three the rest are of little account; yet these three have one common limitation. Johnson was born in 1709, the literary work which made his name belongs, in the main, to the 1750's, and had been rewarded with a pension in 1762, but it was not until the following year that Boswell made his acquaintance. Mrs. Thrale's friendship dates from 1764 or 1765, and Miss Burney, who met him first, and saw him principally, at the Thrales' house, does not make up the trio until 1778. Lives of Johnson were, it is true, written by two men at least who had known him in earlier days, but Murphy's sketch is too brief, Hawkins's conception of a biographer's duties too imperfect, for either to add much of value. Of Johnson's own autobiography only a tantalizing fragment survives:

"The Unknown Johnson" by A. S. F. Gow. From Life and Letters, *VII (September 1931), 200–215. Reprinted by permission of the author. Originally read as a paper for the London Johnson Club, July 1931. In the notes the G. B. Hill edition of the* Life of Johnson *is referred to as B, his* Johnsonian Miscellanies *as Misc., and his edition of Johnson's letters as* Letters. D *means* The Diary and Letters of Madame D'Arblay, *three vols. (London: 1890).*

his letters, when intimate, are of service, but they are intimate only to Mrs. Thrale, and therefore do not help us here.

Of the first fifty-four years of Johnson's life, then, we know little: they occupy less than a quarter of Boswell's book, and almost all we know comes from Boswell. The Johnson with whom we are familiar is a man of settled habits, character, and reputation, the acknowledged head of English literary men, on the threshold of old age, his struggles and most important successes behind him. The unknown child was no doubt father to the man we know, but it is scarcely surprising, Johnson's life having been what it was, that chance-recorded episodes of his earlier years are sometimes hard to fit into the picture. The Johnson who, in the 'teens of the eighteenth century, wore a speckled linen frock and was sick in the Lichfield coach[1] is naturally beyond our vision, but there are other glimpses more disconcertingly remote. Johnson laying in a sword and musket to serve in the militia, Johnson in the '40's walking St. James's Square all night in patriotic conversation with Savage, Johnson sitting in an alehouse with Psalmanazar and a metaphysical tailor, not venturing to contradict[2] —all these put a tax upon the imagination to which it is unequal. Historically, such Johnsons existed, but they have left no clear mark upon our friend, nor are their features discernible in his.

From 1763 onwards, however, the case is different. From that year Johnson stands before us at full length, to be judged not from chance records or from the estimates of others, but by his own words and actions, which, in their consistency and their inconsistency alike, have upon them the stamp of life. The general truth of Boswell's picture is so evident that I need not labour to prove what no one disputes; every reader of the *Life* must have formed for himself a mental picture of its hero tallying very exactly with the summary sketch drawn by its author in the closing pages. But it is not the least of Boswell's merits that the opinions which others entertained of his hero, however different from his own, should yet be intelligible in the light of his account. I am not thinking, of course, of those who saw only one side of Johnson. That Mrs. Boswell and others should think him a bear[3] is nothing, for Boswell's Johnson is bearish. It is nothing that Johnson's friends, Boswell among them, should suppose him the model for Lord Chesterfield's character of the amiable Hottentot,[4] for much in that character really resembles him. But consider Horace Walpole's judgement: "With a lumber of learning," he wrote,[5] "and

[1] *Misc.*, i, 135.
[2] *B.*, iv, 319; i, 164; iii, 444.
[3] *B.*, ii, 66, 269.
[4] *B.*, i, 266.
[5] *Memoirs of the Reign of George III*, iv, 297.

some strong parts, Johnson was an odious and mean character. His manners were sordid, supercilious and brutal: his style ridiculously bombastic and vicious, and, in one word, with all the pedantry he had all the gigantic littleness of a country schoolmaster." The estimate is about as remote from Boswell's as it well could be: "mean" and "little" are the last words to use of Johnson; "lumber of learning" ridiculously inept of a man whose erudition is employed, both in writing and conversation, so appositely and, on the whole, so sparingly as his. Yet Walpole's Johnson is still Boswell's Johnson from another angle, and it is high evidence of the objective truth of Boswell's picture that Walpole's view should seem not merely intelligible, but, in Walpole, natural and even inevitable.

But for all this, even in the record of the last twenty years of Johnson's life, where the information is seemingly so full, it is possible to detect aspects of his character to which Boswell has not done full justice—some because they were shared by Johnson with most of his contemporaries and are characteristic of the age as well as of the man, some because they seldom showed themselves in Boswell's presence, some because Boswell did not appreciate or understand them.

On the first of these classes I will not dwell at length. The general temperament of a people, even where the main lines of character remain stable, varies a little from age to age, and the variation must necessarily dull the vision a little in retrospect. One does not recognize many friends in the novels of the eighteenth century: yet Gibbon called *Tom Jones* "the history of human nature," [6] and Johnson thought *Evelina* worthy of the author of *Clarissa*,[7] and *Clarissa* " the first book in the world for the knowledge it displays of the human heart." [8] From these works you may, at varying prices, supply your Boswell with a useful background. It does not occur, for instance, to a casual reader of Boswell to think of Johnson as a sentimental man, nor did Boswell, it is to be presumed, so conceive him. Yet Johnson, bathed in tears and sobbing at the mere report of a chance-seen grave in a wayside churchyard,[9] or rapturously throwing off coat, hat, and wig to jump again at seventy-two the rail he used to leap as a boy,[10] is the sport of emotions which you or I should hardly feel and certainly should not display.

To Boswell, in short, Johnson was less sentimental than Boswell and not more sentimental than other men. He does not observe the

[6] *Decline and Fall*, Ch. 32.
[7] *D.*, i, 73.
[8] *Misc.*, ii, 251. Cf. *B.*, ii, 174.
[9] *Misc.*, ii, 279.
[10] *Misc.*, ii, 396.

trait because it is not individual. But there is a side to Johnson in
which the sentimental has at least a part where Boswell fails us, not
from lack of perception but from lack of opportunity. It is his attitude
towards women, and especially towards young women, to the stimulus
of whose society Johnson responded with a lightness of heart he seldom
or never showed in men's company. He is half gallant, half fatherly,
sentimental, instructive, charming, grotesque, by turns or as it strikes
you; but he is not Boswell's Johnson. To Boswell, indeed, we owe the
story of the two pretty fools from Staffordshire who had doubts about
Methodism, and were taken to dinner at the Mitre, where one of
them sat on Johnson's knee.[11] But it is an incongruous picture, and
it has lingered in the minds of many readers of Boswell (among them
Rossetti) by reason of its incongruity. Yet if a reader approached
Johnson by way of Fanny Burney he would not think it incongruous.

"If I had no duties," said Johnson once, "and no reference to
futurity, I would spend my life in driving briskly in a post-chaise with
a pretty woman; but she should be one who could understand me
and add something to the conversation." [12] With one half of this
Elysium Johnson was provided by the Thrales, in whose house at
Streatham feminine society, at once intelligent and fashionable, sat
gladly at his feet. Mrs. Thrale herself, who wrote under the smart
of other emotions, does not reveal much of this side of Johnson,
but you will find it in Johnson's letters to her and her daughter or
to Miss Reynolds, and, above all, in the Diary of Fanny Burney, whose
tenderly affectionate relations with Johnson are the most important
supplement to Boswell in all the Johnsonian corpus. Boswell himself
knew little of this (even the dinner at the Mitre was no experience
of his own) for, I suspect, more than one reason. In the first place
Boswell, though admitted, was neither a frequent nor a very welcome
visitor at Streatham, and it was in the inner circle that the mood in
question principally showed itself. In the second, there was that in
Boswell's own character which must necessarily have imposed a check
upon it. Boswell's passions were inflammable, his relations with
women, as Johnson knew, highly irregular, his indiscretion in speech
notorious. It was not by accident that Langton gave him a copy of
The Government of the Tongue, nor that Boswell wrote good resolutions upon its fly-leaf.[13] But Boswell's good resolutions were in the
nature of forlorn hopes, and he was not a man, either by moral
character or social tact, whose presence would encourage Platonic
philandering.

Still, if he had himself seen little of this side of Johnson, he knew,

[11] *B.,* ii, 120.
[12] *B.,* iii, 162.
[13] C. B. Tinker, *Young Boswell,* p. 15.

nevertheless, that it existed. "He would not," he told Hannah More, "make a tiger a cat to please anyone." [14] But he was, for all that, well aware that his tiger had a kittenish side, and if that side is not represented in the *Life* the fault is not Boswell's. Twice in the year 1790 he had caught Fanny Burney outside St. George's Chapel at Windsor and begged her to communicate some of Johnson's letters. "I want to show him," he said, "in a new light. Grave Sam and great Sam and solemn Sam and learned Sam—all these he has appeared over and over. Now I want to entwine a wreath of the graces across his brow; I want to show him as gay Sam, agreeable Sam, pleasant Sam; so you must help me with some of his beautiful billets to yourself." [15] But Miss Burney had scruples, Boswell's Johnson remains ungarlanded, and you may lay down the *Life* without discovering why on Dec. 20th, 1784, one young lady at least could not keep her eyes dry all day.[16]

There is, however, another side of Johnson which Boswell knew, and has, indeed, touched upon, but to which he has, I think, deliberately done less than justice—I mean Johnson's capacity for boisterous, open-hearted, irrational laughter. The supreme instance of this trait we certainly owe to Boswell. It is that familiar night in the Temple when Johnson encountered in the rooms of his lawyer the gentleman had just drawn up a will bequeathing his estate to his sisters. Johnson, amid the growing discomfort of the audience, subjected him to merciless ridicule and uproarious laughter which culminated after Johnson and Boswell had left the party and emerged from the Temple, when Johnson "burst into such a fit of laughter that he appeared to be almost in a convulsion; and, in order to support himself, laid hold of one of the posts at the side of the foot pavement, and sent forth peals so loud, that in the silence of the night his voice seemed to resound from Temple-bar to Fleet-ditch." [17]

That is the only full-length of such a scene, but incidents of the kind do not occur only once in a man's life, and, as was to be expected, there are traces of others. There was, for instance, Mr. Coulson, about to take a country living, whom Johnson chose, with uproarious merriment, to fancy as a future archdeacon;[18] and even in Boswell echoes of similar outbursts may be heard. But these flights of fantastic absurdity and helpless irrational laughter belong to larger natures than Boswell's, and to nations less cautious than his to count the cost. Boswell himself could appreciate wit, and he could sit giggling in a corner if the scene staged before him degenerated for a moment

[14] *Misc.*, ii, 206.
[15] *D.*, ii, 378.
[16] *D.*, i, 288.
[17] *B.*, ii, 262.
[18] *Letters*, i, 325.

into farce: but humour was not his line, and humour running riot plainly puzzled and disturbed him. Sometimes, no doubt, he made no record of it. Of one day in 1775 he writes: "I passed many hours with him of which all my memorial is 'much laughing'." [19] Even the scene in the Temple he records with some apology—it is "here preserved that my readers may be acquainted even with the slightest occasional characteristicks of so eminent a man." [20]

But when Boswell describes such scenes as slight and occasional characteristics he misleads you, for Johnson's other friends form a cloud of incontrovertible witness to the contrary. Murphy said "he was incomparable at buffoonery"; Miss Burney that "he has more fun and comical humour and love of nonsense about him than almost anybody I ever saw"; Hawkins, Mrs. Thrale, and Garrick all speak to the same effect. "Rabelais," said Garrick, "and all the other wits are nothing to him. You may be diverted by them; but Johnson gives you a forcible hug, and shakes laughter out of you, whether you will or no." [21] And since Garrick said this to Boswell, who has recorded it in his book, to the surprise of attentive readers, Boswell must stand arraigned for the fact that one of Johnson's most striking characteristics is all but absent from the *Life*.

No doubt here also there are extenuating circumstances. There are recorded a number of occasions when Johnson gave his exuberance of spirit full play. We see him pursuing the college servitor at Pembroke with the noise of pots and candlesticks to the tune of "Chevy Chase," roused at three in the morning to frisk in Covent Garden, exchanging coarse badinage with a bargee, touching off damp fireworks in Marylebone Gardens, swarming trees, or kicking off slippers to race with young ladies on the lawn.[22] But it is worth notice that, though half these incidents are recorded by Boswell, none are from his own experience; and it is hard to resist the conclusion that here also Boswell's presence imposed some restraint upon Johnson. And I do not think it very hard to guess why. "It is surely better," Johnson had once written, "that caprice, obstinacy, frolick and folly . . . should be silently forgotten, than that, by wanton merriment and unseasonable detection, a pang should be given to a widow, a daughter, a brother or a friend." [23] But both in writing and in conversation he had often urged the necessity of frankness in biography.[24] He knew,

[19] *B.*, ii, 378.
[20] *B.*, ii, 262.
[21] *B.*, ii, 231, 262n. Birkbeck Hill has already criticized Boswell's omissions on this head.
[22] Hawkins, *Life*. Ed. 2, p. 13. *B.*, i, 250; iv, 26, 324. *Misc.*, ii, 278.
[23] *Lives of the Poets*. Ed. 1818, ii, 255.
[24] *B.*, i, 30; iii, 155; v, 240.

moreover, from 1772 at least, that Boswell intended to write his life,[25] and, since, a year later, he had read Boswell's Journal of the Scottish tour, he was aware of the kind of materials Boswell was collecting. Johnson was a man above petty vanity, but with Boswell constantly prying into his correspondence or eavesdropping behind his chair he had little chance of forgetting either the task Boswell had undertaken or the lack of discretion he might bring to it.[26] "One would think, the man had been hired to be a spy upon me," he once wrote,[27] and it is natural that such a presence should have imposed some restraint upon him.

There was, however, another motive which may well have contributed to the same end. Boswell habitually thought and often spoke of Johnson as his "guide, philosopher and friend," [28] and Johnson, accepting complacently enough the post of Mentor, had never been backward either in advice or in reproof. It is natural that he should have reinforced precept with example, and refrained in Boswell's presence from caprices which transgressed the ordinary conventions of society. The exuberance of spirit which led Johnson, at fifty-five, to empty his pockets and roll down a hill was too like that which led Boswell to moo like a cow in the theatre, or to sing a song of his own composition six times through at a Guildhall banquet, for Johnson to display it unnecessarily before a man who, by his own just estimate, was usually in too high spirits or too low.[29]

But granting that Boswell saw less of this side of Johnson than, let us say, Garrick, I still cannot help suspecting him here of some lack of candour. The fact is that Boswell thought uncontrollable amusement vulgar, and had even said so in print. "Whether loud laughter in general society," he had written in his *Tour to Corsica*,[30] "be a sign of weakness or rusticity I cannot say; but I have remarked that real great men, and men of finished behaviour, seldom fall into it." That Johnson was a great man, Boswell, with good reason, never doubted. That he was a man of finished behaviour, no one except Johnson has ever supposed. But Boswell, who, in his peculiar way, was a snob, may not have cared to rub it in, and I have sometimes thought him uneasy about this aspect of his hero. It is, at any rate, significant that when Dean Barnard bluntly said that Johnson was not a gentleman, Boswell, though he recorded the observation in his notebook, was careful to exclude it from the *Life*.[31]

[25] *B.*, ii, 166, 217.
[26] *B.*, v. 364. *Misc.*, i, 175; ii, 389. D'Arblay, *Memoirs of Dr. Burney*, ii, 193.
[27] *Letters*, i, 330.
[28] *B.*, iii, 6; iv, 122, 420.
[29] *Misc.*, ii, 391. *B.*, v, 396. C. Rogers, *Boswelliana*, pp. 330, 283.
[30] Ed. 1923, p. 37.
[31] *Boswell's Note Book.* (1925.) p. 18.

But the upshot of all this is perhaps no great matter. Boswell's portrait, one might say, is not the less true because some of the features are in shadow. And if chance has left other sketches in which those features appear, it may be amusing but it is not important to display them by the side of the finished picture, since the latter provokes no feeling of incompleteness. And this is, I think, true of all the aspects of Johnson I have so far touched upon. But there are still subjects upon which Boswell really does leave me with a sense of incompleteness, where neither he nor the other sources persuade me that we have the whole truth before us. How came a man of Johnson's principles and intellect to lead the stagnant life Johnson is depicted as leading? How, when all allowance is made for Boswell's good qualities, could Johnson really tolerate him? These are questions to which some answers, though not, I think, complete ones, can be given. I pass them over to put the more fundamental one: to what is Johnson's domination over his own circle ultimately due? That domination is the most striking thing about Johnson, yet Boswell throughout assumes it as a matter of course, and his reader, unless he too assumes it, is scarcely helped at any point to understand it.

Consider the facts for a moment. Goldsmith once charged Boswell with making a monarchy of what should have been a republic,[32] and it is true that Boswell was quick to resent interference with Johnson's sovereign authority. But if Boswell did not interfere, Johnson himself loosed the thunders soon enough, and the king was little less royalist than his minister. One instance, and that a famous one, will suffice. The focus of the Johnsonian circle was the Club, which in 1776 included, besides Burke, Reynolds, Gibbon and Sheridan, Forbes, Colman, Warton, Barnard and Chamier, all men of note in their day. All of these wished Johnson to reconsider the Latin epitaph he had written for Goldsmith, their personal friend; yet, though the request was reasonable, none dared make it in person. They addressed therefore to Johnson a round-robin, couched in the most deferential terms. Johnson received it with good humour and paid not the faintest attention to its contents.[33] Now these men were all of them, it is true, younger than Johnson, yet some were his superiors in genius, most in status, all in wealth and what the world counts success. What explains his or their behaviour on this occasion? Why did Johnson exact, or his friends readily pay, such exaggerated deference?

Let us remember how Johnson's friends were in account with him. On the debit side there was, first, some glory in knowing intimately the author of *The Rambler, Rasselas,* and the Dictionary. It was not, perhaps, even then, a principal motive, but it was the impulse which

[32] *B.,* ii, 257.
[33] *B.,* iii, 83.

brought Boswell into the circle, and it must be allowed some weight. Secondly, there was the fundamental soundness and sanity of Johnson's character. It is the most striking thing about Johnson, and to the lighter-headed of his friends—Boswell, Goldsmith, and perhaps Garrick—it may well have meant much. "I looked at him," says Boswell, "as a man whose head is turning giddy at sea looks at a rock or any fixed object." [34] And finally, there was his conversation. In these unconversational days, it is hard even for those who know and enjoy it best to do full justice to its powers; but it had, it is plain, an irresistible attraction for his contemporaries, to which those who were in his company only once bear as enthusiastic witness as do his intimates. "I verily think," says Richard Cumberland, "he was unrivalled both in the brilliance of his wit, the flow of his humour, and the energy of his language"; and Dr. Lettsom, more briefly: "he had a heavy look, but when he spoke it was like lightning out of a dark cloud." [35]

But, however spellbound Johnson's conversation might hold you, it had its disadvantages. For one thing, it was of the didactic kind. "He would take up a topic," says one acquaintance, "and utter upon it a number of *The Rambler*." "It is almost impossible to argue with him, he is so sententious and so knowing," says another.[36] But if argument did arise, Johnson was apt to argue for victory rather than truth, and, in Goldsmith's familiar phrase, knocked you down with the butt if his pistol missed fire. The extravagant rudeness of his dialectic is well known. "Pray, Sir, what you are going to say, let it be better worth the hearing than what you have already said"; or "I would advise you, Sir, never to relate this story again: you really can scarce imagine how very poor a figure you make in the telling of it." [37] And if Johnson took it into his head to tax you with being drunk, you might be as intemperate as Boswell or as sober as Reynolds for all the difference it made. "You may observe," said Johnson with complacency, "that I am well-bred to a degree of needless scrupulosity";[38] and gross unwarrantable insult is not made more palatable by the assumption of superior breeding.

And to Johnson's deliberate rudeness we may add that he was shortsighted, grossly careless of dress, coarse in table manners, and that his unconscious tricks and mannerisms attracted notice in public, and, in private, more than once caused him to be taken for a lunatic.[39]

[34] *B.*, v, 154.
[35] *Misc.*, ii, 76, 402.
[36] *Misc.*, ii, 391, 401.
[37] *Misc.*, i, 242; ii, 258.
[38] *Misc.*, i, 169; cf. ii, 260, *B.*, v, 363.
[39] *B.*, i, 145. *Misc.*, ii, 275, 297, 400, 424.

He was, in short, a remarkably unpresentable figure, and Fanny Burney has drawn a gloomy picture of the terror his presence spread in the society at Brighton.[40] Mrs. Thrale, who was a shrewd woman, and viewed her acquaintances, even those she liked, with curious detachment, once made a marksheet for her friends in which 20 was full marks for each subject.[41] Johnson got full marks for Religion, Morality, and General Knowledge, 19 for Scholarship, 16 for Humour, 15 for Wit: for Person and Voice, Manners, and Good Humour he got 0. The marking is evidently not far astray; yet good humour makes more and firmer friends than religion, morality, or even general knowledge, and no man had more devoted friends than Johnson.

From this brief reckoning then appears what strikes me continually when I read Boswell. The known causes do not adequately account for the known results. Boswell's Johnson is a great and dominating figure with many admirable and some lovable qualities, yet you do not feel that you would have constantly sought his society or thought it more than worth the price in deference and humiliation at which it was to be attained. Some essential but volatile element of that potent spirit has escaped in the bottling. Boswell has caught and conveyed to us an overpowering personality, but he has not conveyed, save by the repeated statement of its effects, some quality of it which is now beyond recovery. We know much, but we shall never know all, that Johnson's friends saw in him.

It happens that the second great English biography is also that of a man who, though in a very different way, exercised a great fascination over his friends. Yet Lockhart's *Life of Scott* leaves in the reader's mind no such dim questioning as this. From the first page, Scott stands out as an eminently lovable man, and the qualities by which he bound his friends to him may be seen in this guise or that on every page. Now the moral is certainly not that Lockhart knew his business better than Boswell. Time has fully justified the opinion which, without undue modesty, Boswell himself expressed. "I am absolutely certain," he wrote, "that my mode of biography . . . is the most perfect that can be conceived";[42] and Lockhart's claim to the second place among biographers is based upon the success with which he pursues Boswell's methods, not upon his departures from them. If Lockhart, then, sets down, as I think he does, the whole of Scott, and Boswell's Johnson is incomplete, the cause is to be looked for in the subjects, not in the authors, of the two books. Scott's transparent simplicity has enabled the lesser artist to make his work complete:

[40] *D.*, i, 240.
[41] C. Hughes, *Thraliana*, p. 21.
[42] *Letters of James Boswell.* Ed. C. B. Tinker, p. 344, cf. p. 342, *B.* i, 30.

Boswell's jewel was a crystal of more clouded substance and more complex structure, and some of its facets are in his picture imperfectly defined. Yet Boswell devoted to Johnson such study as surely no man ever before or since has devoted to another. "Every look and movement," says Fanny Burney, "displayed either intentional or involuntary imitation. . . . His heart, almost even to idolatry, was in his reverence of Dr. Johnson." [43] And the moral which seems to emerge is this. The Johnson whom Boswell has drawn is evidently a great man, but the incompleteness of his biography is still more conclusive proof of that greatness and the final vindication of Boswell's choice of subject.

[43] *Memoirs of Dr. Burney,* ii, 191.

Boswell: The Life of Johnson

Ralph H. Isham, Joseph Wood Krutch, and Mark Van Doren

Van Doren: Gentlemen, we have as a subject for our conversation today the biography which most people agree is the greatest of all biographies. A good old question about Boswell's *Life of Johnson*—a question that was asked at the very beginning, in the late eighteenth century, and that Macaulay thought he answered finally in the nineteenth century—is this: whether the greatness of the book is to be explained by the nature of its subject or by the nature of its author. Is it Johnson himself, he having been what he was, who makes this book so great, or is it the artistry of Boswell? Mr. Krutch, have you an answer offhand?

Krutch: I have only the answer that it was a very happy conjunction of two things. Macaulay's famous paradox that Boswell wrote a great book because he was a fool is, of course, absurd; folly cannot be great. Boswell often did foolish things, very foolish things. He often said foolish things. But it was his wisdom, not his folly, that made the *Life of Johnson* a great book.

Van Doren: Colonel Isham, since you possess the wonderful collection known as the Boswell Papers you might have a special answer. Or do you think it's an outmoded question?

Isham: The question is often asked: which is the greater man? Certainly one must say that Johnson was the greater man: greater in caliber, greater in learning, greater in philosophy. But I think one can safely say that Boswell was the greater genius. He had a great, unconscious intellectual talent, which one finds, for example, in the fact that in his journal, which was his confessional as well as his repository for the people he collected—

Van Doren: It is one of the items among your papers.

"Boswell: The Life of Johnson" by *Ralph H. Isham, Joseph Wood Krutch, and Mark Van Doren. From "Boswell: The Life of Johnson"* (a recorded radio program) in New Invitations to Learning, ed. Mark Van Doren. (New York: Random House, Inc., 1942), pp. 285–96. Reprinted by permission of Mark Van Doren.

Isham: Yes, we have his notes or journal from 1761 to 1794, just before he died. In it he wrote so frankly, so honestly, and so simply, that when he came to write the *Life of Johnson,* perhaps on fifteen occasions he tore pages out of this journal and, with some minor corrections, sent them to the printer as printer's manuscript. I think that's a testimony of his unconscious intellectual talents.

Van Doren: These were pages that he had written immediately after conversations?

Isham: Very shortly after, yes.

Van Doren: His habit was, if he had been with Johnson, to go home as quickly as possible, I take it, and set down what had been said.

Isham: Well, not always as quickly as possible, because Boswell liked a good time, and if he was having a good time, he didn't go home early. But as soon as he regained consciousness, shall we say, he made at least short, rough notes of these things—anyway, from one day to a few weeks after—and then expanded them into his journals without any sacrifice of accuracy.

Krutch: Nevertheless, the fact remains that when we are reading Boswell's *Johnson* it is Johnson that we are aware of. That is one of the proofs, perhaps, of Boswell's genius. He knows that we are interested in Johnson, not in Boswell; it is part of his art to retire himself from the pages so that we seem to be coming in almost direct contact with Johnson. It isn't really direct. If it were, it wouldn't be so interesting or so pointed; but Boswell creates the illusion that he is of little importance, of less importance than he really is.

Van Doren: And that is a very profound tribute to him, as it would be to any artist. One reason, I suppose, that we adore Shakespeare is that he never makes us think of him while we are reading one of his plays. Boswell does this very subtly—I imagine you would agree, Mr. Krutch—because sometimes he does it by putting himself apparently in the foreground, but only as an object from which our glance can be thrown to Johnson himself.

Krutch: He admitted, of course, that he was perfectly willing to appear as a fool in the pages, if by appearing as a fool he could draw from Johnson one of the remarks that delighted the reader.

Van Doren: There is a very interesting paragraph, at the conclusion of his introduction, in which he says that he is not going to do that quite as often as he had in his *Tour to the Hebrides.* The world had misunderstood him, and the world was a fool; it had not been able to recognize that it was artistry oftentimes which compelled him so to behave.

Isham: Well, sir, Boswell was always willing to take a rebuff from Johnson, and Johnson's rebuffs were wholehearted.

Van Doren: Yes, they were.

Isham: To draw him out, you see. And I can give an example of that. He writes in the *Life*—he does not disclose that it was himself; he just says "a gentleman," but it *was* Boswell, we find from the *Journal*—he writes: "A gentleman, using some of the usual arguments for drinking, added this: 'You know, sir, drinking drives away care and makes us forget whatever is disagreeable. Would you not allow a man to drink for that reason?' Johnson: 'Yes, sir, if he sat next to you.'"

Van Doren: You mean next to Boswell?

Isham: Boswell. But he does not disclose it in the *Life*. He was always glad to take a rebuff if it drew the old boy out.

Van Doren: But on another occasion he might very well have made it clear that it was himself, if to do so had served an artistic purpose.

Isham: An *artistic* purpose, yes.

Krutch: Now, those who go to the opposite extreme and think Boswell completely a genius rather than a fool sometimes speak as though Dr. Johnson were Boswell's creation. It is often said that Johnson is remembered only because of Boswell. I think that is absurd for several reasons. One of them is that, after all, various other people did leave accounts of Johnson. Not one of these is so good, not one of these is so vivid, as the Johnson of Mrs. Thrale or of Fanny Burney. But all of them present the same Johnson. In other words, Boswell's *Johnson* is more of a portrait than a creation.

Van Doren: I dare say it is. That merely brings up the metaphysical question where, in the case of any portrait, say by a great painter, credit lies for the greatness of the result: in the subject who had a soul to be discovered or in the man who was capable of discovering that soul? It is a question to which I suppose you never find the answer.

Krutch: Boswell's Johnson is more continuously and perfectly Johnson than Johnson ever was himself, but it is still the essence of Johnson, not something else.

Van Doren: Boswell had a great capacity for recognizing interesting people and for devoting himself to them, not at all as a toady or a sycophant, I take it, but as one whom, incidentally, they welcomed in their presence. It is important to remember, when we hear it said that Boswell was the toady of Johnson, that Johnson almost from the beginning liked Boswell, and a short time after he met him insisted on accompanying him to Harwich when he was going to the Continent.

Isham: Yes, sir. And Boswell recorded it with, I imagine, great satisfaction. He says as his boat drew away from Harwich that he watched Johnson standing on the shore rolling his great bulk until finally he turned and disappeared into the town.

Krutch: It is worth remembering that one of the things Johnson scolded Boswell for was Boswell's continual demand that Johnson should reassure him concerning his feelings for Boswell. On one occasion I remember Johnson said: "I love you. Write that down in your notebook and don't ask me about it again."

Isham: I can explain that, I think, psychologically, sir. Boswell was always suffering from hypochondria; he had what we would call today a great inferiority complex, and he needed Johnson. Johnson was his strength, almost his religion. The strong moral philosophy of Johnson saved Boswell from his weaknesses, undoubtedly, and you find very often in his *Journal:* "Be like Johnson. Remember you are his friend." That was a constant strength to him; he worshipped Johnson. The theory that he was a sycophant, that he was just trying to shine in reflected glory, is not true—he really worshipped Johnson. You find that before he ever meets him he is recording in his *Journal* how he is strengthened by reading *The Rambler.*

Van Doren: As a matter of fact, that was the reason he wanted to meet Johnson, wasn't it?

Isham: It was the chief reason.

Van Doren: And not the purely professional reason, either, that he some day wanted to write a biography of this man.

Isham: Exactly, sir. It was as if a man wanted to meet the god he believed in.

Van Doren: That accounts, I think, for the extremely dramatic character of those pages in which Boswell, who up till now has been writing a biography of a man he had not known, describes his first meeting with him. It is almost as if an annunciation, a visitation, were occurring. The front door of a shop opens and in comes, at last, the great man. It is very exciting.

Krutch: I remember also that Boswell in his own *Journals* rather frequently asks whether or not something he has done or is about to do is worthy of James Boswell—which is, of course, the vanity of a man who distrusts himself rather than the vanity of a man who is sure of himself.

Isham: I find that an excellent observation. . . .

Van Doren: One thing I admire Boswell for is his recognition of Johnson's great gift for what Boswell himself called imagery. Johnson, of course, was a wonderful talker, and many people make the mistake of supposing that he always talked in polysyllables, in abstractions. The conversation of Johnson as here reported is most of the time, as a matter of fact, made up of short words and fascinating words. He had a wonderful gift for summing up a thing he wanted to say in terms of things easily seen. There is the occasion, for instance, on which Boswell was asking him, as he often had before, why Johnson was so grudging in his praise of Robertson's history. Johnson said: The man wraps his gold in wool; most of the space in his box is taken up by the wool. Now I think that an incomparably brilliant way of saying what Johnson wanted to say. Wouldn't you agree that Johnson generally talks that way in this book?

Isham: Johnson probably spoke in a much more pedantic fashion than Boswell records. You'll find that in Scotland Boswell has to defend Johnson for his use of large words by saying that he was so long a school teacher—teaching Latin, so that it became second nature to use Latin words. I think he probably—in fact, we have evidence—tones the big words down and shortens them up a bit.

Van Doren: He was, of course, right in doing so.

Krutch: There is also evidence in the notebook, as I remember, that Boswell sometimes made Johnson talk better than Johnson did; that is to say, he edited the conversation. But to come back again to the question in what way Johnson was a great man. We've touched on one of the things that ought to be said. Carlyle talked about the hero as writer, the hero as soldier, and so on. You would have to call Johnson the hero as talker, because, though Johnson was a good writer, there is no doubt about the fact that the thing he did best was talk.

Van Doren: That is what Boswell himself says.

Krutch: Yes.

Isham: I think one thing must be said of Boswell—he was the greatest reporter of all time. He had a great sense of accuracy; he was a Scot; and also he was a lawyer—I'm speaking of the eighteenth century. He had doubly, therefore, a sense of truth and accuracy. And he also was the inventor of the interview, of personal journalism. Perhaps if he lived today he would have millions and be in danger of getting a peerage.

Van Doren: By the way, there's a certain irony there. I happen to remember a moment fairly late in Johnson's life when, after someone had heard him praise a certain lady whom he didn't know—I think an actress—he was asked why he didn't go to see her. He said: Sir, the reason I do not go to see her is that these days everything like that gets in the newspapers. And at that very moment, as you say, the inventor of the interview, the greatest reporter of all times, was preparing the greatest of all personal lives.

Isham: And never doubted his ability artistically to do it. Boswell often doubted his ability to stick at it long enough to complete it, but his artistic ability he never doubted for one minute.

Van Doren: Mr. Krutch, you were saying that Johnson's essential greatness lay in his power of conversation. Was there any special thing that he preferred to talk about, and do you have in mind any special thing that he said?

Krutch: There is no special subject that he talked about. Yet Johnson and Boswell, different as they were in temperament, were alike in one respect: both were interested primarily in men and manner. That is the reason Johnson liked London and didn't like scenery. The subject of conversation for Johnson, as for so much of the eighteenth century, was what human nature is like, how people behave. . . .

Krutch: Johnson was a man who, of course, loved books. But I think he could have got along without books, whereas he couldn't have got along without conversation. It was *the* necessity of life as far as he was concerned.

Van Doren: Johnson hated to be alone, didn't he?

Krutch: Yes, his friends sometimes hesitated to go see him, because it was so hard to get away. He couldn't bear to be left at night.

Van Doren: And he had one of the most charming of traits—a thing that always makes us love our friends, and probably is the reason we choose those particular friends. When you went to call on him and to suggest that he go somewhere, for dinner or for a late supper or what-not, he was always willing to go. He would drop everything, and he was usually more frisky than the rest. Remember that occasion on which he bounded out of his lodgings with the remark: "I have a mind to frisk with you this evening, gentlemen."

Krutch: No one has mentioned the fact that Dr. Johnson was a pessimist who enjoyed life. Theoretically, he was a pessimist; practically, he was a man beset by fears, by illness, and by gloom. And yet few men ever lived who enjoyed life more thoroughly. He was a moralist without being an ascetic. Those pleasures—of society as well as of literature—which he could enjoy, he savored to the full.

Van Doren: Mr. Krutch, I seem to remember his saying—you could correct this quotation, or Colonel Isham could—that he felt life was something to be endured rather than enjoyed. That fits in, does it, with what you are saying?

Krutch: "There are more things in life to be endured than to be enjoyed." But he didn't say there is nothing in life to be enjoyed. On the contrary, he was a man who found a very large number of things to be enjoyed, and he enjoyed them very thoroughly.

Van Doren: His true greatness as a talker for me is that he was willing and able to talk about anything under the sun; no subject ever arose, no matter how little or how great, but what he immediately had a store of things to say about it.

Isham: He had a great store of knowledge, and I think he perhaps spoke well on a subject even if he had no knowledge of it. He would always speak with great authority.

Van Doren: With great authority. And even when he had no knowledge he had a kind of wisdom which told him what the limits of the subject were, what the bearing of that subject was on other subjects.

Isham: Great wisdom. And, of course, one of the great things of this book to me is that in it he reveals us to ourselves a bit. We are all subject to the same hopes, fears, doubts, and here we learn how a man gets through them, faces them, experiences them. That is valuable for us. . . .

Reflections on a Literary Anniversary

by Donald J. Greene

It used to be maintained, of course, that Johnson's writings are well lost for the sake of having in our possession Boswell's *Life of Johnson,* "the greatest biography ever written." This was one of Macaulay's journalistic *tours de force:* "Homer is not more decidedly the first of heroic poets, Shakespeare is not more decidedly the first of dramatists . . . than Boswell is the first of biographers. . . . Eclipse" —a race-horse—"is first, and the rest nowhere." When Macaulay states something in this vein of ludicrous exaggeration, it is as a rule nonsense. For, strictly, Boswell's book can hardly be seriously termed a biography at all. It is a series of excerpts from his huge diary, those dealing with the times, during the latter two decades of Johnson's life, when he was in Johnson's company (on a total of 425 days, someone has worked out, a fourth of them during the Hebrides trip). It is introduced by a most inadequate summary, obtained at second-hand and often very inaccurate, of the first fifty-five years of Johnson's life, and patched together by even more inadequate summaries of the periods, sometimes two or three years at a stretch, when Boswell did not see him (for Boswell lived in Scotland and came to London only occasionally). What are we to call a book in which we read, "During this year there was a total cessation of all correspondence between Dr. Johnson and me . . . ; and as I was not in London, I had no opportunity of enjoying his company and recording his conversation. To supply this blank, I shall present my readers with some *Collectanea"* —miscellaneous anecdotes, of no particular date—"obligingly furnished to me." A biography? Surely not. It is an edited diary.

As a diary—a record, essentially, of *Boswell*—the *Life* is a work of art, a minor masterpiece. As a biography—a serious attempt to set down in coherent order the significant facts of a person's life and to make such sense of them as the writer's lights afford—Sir John Haw-

"Reflections on a Literary Anniversary" by Donald J. Greene. Excerpted from Queen's Quarterly, LXX (Summer 1963), 198–208. Reprinted by permission of the author and Queen's Quarterly. [Begins and ends with an imaginary scene in Tom Davies' back parlor in May 1763, different from the one described by Boswell.]

97

kins's *Life of Johnson,* recently rescued from nearly two centuries of
oblivion by Bertram Davis, comes closer to the ideal. Hawkins was
better qualified for the task than Boswell. He was better educated,
more experienced, probably more intelligent than Boswell. He was a
distinguished magistrate in London for twenty years and an important
pioneer musicologist, Boswell never more than a strikingly unsuccess-
ful practitioner at the bar. Although one still encounters the preposterous assertion that if it had not been for Boswell no one would now
hear anything of Johnson (another of Macaulay's brain waves), the
truth is just the reverse: if it had not been for his connection with
Johnson, Boswell would never have emerged from the crowded limbo
of minor eighteenth-century scribblers. His writings on Corsica might
have rated a footnote in historical treatises; when his journals were
discovered, a one-volume selection from them might have been issued
by some enterprising publisher—and not improbably remaindered.
Nothing else he wrote without Johnson in it is worth noticing. Most
important, Hawkins was of Johnson's own generation and a similar
background. They were struggling young writers together on the
Gentleman's Magazine before Boswell was born. They were Londoners
together, while Boswell was pursuing his amours on the slopes of
Castle Rock in distant Edinburgh. At the end, Hawkins attended
Johnson through the physical and spiritual agonies of his last years,
was at his death-bed, and was appointed by Johnson his chief executor.
In the last three years of Johnson's life, Boswell was in London a
total of four months, the last time six months before Johnson's death,
and was not even mentioned in his will. Hawkins had a clear understanding of Johnson's religious and political views, and provides excellent discussions of them, which students of Johnson would do well
to master. Boswell, a Scot, and a Presbyterian by upbringing (with a
youthful excursion into Roman Catholicism), understood neither, and
covers both subjects with confusion.

Only recently scholars have begun, somewhat reluctantly, to acknowledge how much serious distortion there is in Boswell's *Life.*
There is, for example, his treatment of Johnson's marriage—to fantastic old Tetty, who, at the time of her death in 1752, was sixty-three
and somewhat of an alcoholic and opium-addict, while her husband
was in his vigorous early forties. It suits Boswell's purpose to exhibit
Johnson as maintaining an unwavering idyllic devotion to her. Hawkins had taken a rather more realistic approach. Boswell, leading up
to an attack on "a dark and uncharitable assertion" by Hawkins,
pronounces, "That his love for his wife was of the most ardent kind,
and, during the long period of fifty years, was unimpaired by the
lapse of time, is evident from the various passages in the series of
his *Prayers and Meditations*"; and he quotes some suitably tender

passages from them, and goes on to tell how Johnson preserved her wedding-ring in a little round wooden box—it might be a nineteenth-century official biographer indignantly repelling the suggestion that Queen Victoria ever wavered in her devotion to Albert.

What Boswell omits to mention is that among the passages he had transcribed from Johnson's private "Prayers and Meditations" was one, a year after Tetty's death, which begins, "As I purpose to try on Monday to seek a new wife"—a fairly important piece of information, one would have supposed. But Boswell in his wisdom decides that it is better for his readers not to know about it. James L. Clifford, in his *Young Sam Johnson,* describes briefly another scrap of paper in Boswell's hand, a memorandum of an interview with Elizabeth Desmoulins, in which she tells how, during Tetty's last years, Johnson would sometimes call her into his bedroom—she was in her early thirties—make her sit on his bed, and kiss and fondle her ardently, though never going any farther. So it was not quite a Victoria-Albert relationship after all. This last note of Boswell's he labelled *Tacenda* —things to be hushed up. Why he wanted to hush them up is a subject for speculation. Was he simply incurably conventionally minded? Did he feel a little guilty about his own compulsive infidelity to poor Margaret Boswell? Was he perhaps reluctant to make Johnson seem a little *too* human?

In another matter, to be sure, Boswell is only following the lead of an earlier Johnsonian editor, the Reverend George Strahan, when he omits from the solemn prayer Johnson composed on his death-bed for his last communion, the important petition "Forgive and accept my late conversion." A Great Cham who had undergone the vulgar Evangelical experience of "conversion"? Perish the thought. The authentic version of the prayer is available to us, as it was to Boswell, in Hawkins's *Life.* But significantly, in the numerous collections of Johnson's prayers that were later printed, it is always the Strahan-Boswell bowdlerization that is presented, with the result that Johnson is thought by many to be more "High Church" than he actually seems to have been—which was no doubt what Strahan and Boswell intended should happen. Johnson's defences of Roman Catholicism are reported in the *Life* in appreciative detail; an attack on it (October 12, 1779) is skimmed over in two unmemorable sentences: "He this evening expressed himself strongly against the Roman Catholicks, observing, 'In everything in which they differ from us they are wrong.' He was even against the invocation of Saints; in short, he was in the humour of opposition." Boswell did his work so well that not long ago some eminent scholars were asking for help in locating the place in the *Life* where Johnson says of the *Presbyterians,* "In everything in which they differ from us they are wrong." They had searched all the passages dealing with Presbyterianism, and unaccountably could not find it.

As for Johnson's political views, Boswell does his best—and his best is pretty good—to make Johnson appear a sentimental, Jacobitic, Romantic Tory, of the Sir Walter Scott kind, like Boswell himself, instead of the tough, skeptical, down-to-earth practical conservative that he was. He does this by insinuating his own gushings into his account of Johnson, so that the reader has to be fairly alert not to incorporate them into his picture of the older man. "The infant Hercules of Toryism" he labels Johnson, telling a quite incredible story of how, when he was less than three, the baby insisted on being taken to hear the Tory Dr. Sacheverell preach in Lichfield. When he mentions Johnson's *Life of Waller,* he comments on "how nobly Johnson might have executed a Tory history of his country"; and much else of this kind. When Johnson says something about politics, Boswell may edit it heavily and unscrupulously. In his original journal he had noted a remark by Johnson, expounding the well-known legal principle that the Crown is exempt from prosecution, in these words:

> Johnson showed that in our constitution the King is the head, and that there is no power by which he can be tried; and therefore it is that redress is always to be had against oppression by punishing the immediate agents.

In the *Life* he transforms it into a fervent profession of monarchism:

> JOHNSON. Sir, you are to consider, that in our constitution, according to its true principles, the King is the Head; he is supreme; he is above every thing, and there is no power by which he can be tried. Therefore it is, Sir, that we hold the King can do no wrong; that whatever may happen to be wrong in government may not be above our reach, by being ascribed to Majesty. Redress is always to be had against oppression, by punishing the immediate agents.

"He is supreme," "he is above everything," "being ascribed to Majesty" seem to be pure Boswell, grafted on to Johnson. How much of this kind of editing, or rather gratuitous falsification, the text of the *Life* contains awaits serious study.

The result of all this sort of thing is to make Johnson a much simpler person than, in his complexity, he really was—simple-minded in his uncritical devotion to his wife, the Church, the monarchy; simple-minded, lovable, and, in the end, slightly ridiculous. Could Boswell have intended just this outcome? It sounds preposterous: the name of Boswell has become proverbial for whole-hearted, all-embracing devotion to an older and more eminent person. Still——. Bruno Bettelheim reviewing Ernest Jones's biography of Freud recently, called attention to the time-honoured tradition of the disciple subtly undercutting the master, pointing out (in the most reverent way) his little imperfections, bringing him down to the disciple's size or a little lower, making his teachings comprehensible to the masses by diluting them with the disciple's—as, Bettelheim complains, St. Paul did for Jesus. Perhaps he might have also cited Boswell on Johnson. The psychology is easy enough to understand, in terms of what we now know about father-figures. Boswell had been denied all affection by his father, the austere Calvinist judge Lord Auchinleck—witness the letter printed at the end of Boswell's *London Journal,* where he takes young Jamie's character to pieces with cold, grim, systematic relish. So he spent his life in a desperate search for a substitute, running about Europe after "great men"—Lord Keith, Voltaire, Rousseau—to whom to attach himself. At last Johnson took pity on him.

But the person you have to depend on emotionally you come to resent, whether you admit it or not. Why, to take a seemingly small point, did Boswell go through the proofs of his first book about Johnson, the *Hebrides Journal,* and carefully change all the "Mister Johnsons" to "Doctor Johnsons"? Johnson was no more in the habit of speaking or thinking of himself as "Doctor Johnson" than any other sensible holder of an honorary doctorate. But thanks to Boswell, the average reader is probably convinced that he did; and so he seems stupidly pompous. Did Boswell not suspect that this would be the result? He gives us some interesting examples of what may be called the technique of damnation by inadequate defence. In his account of Johnson's *Dictionary,* after some paragraphs of rather formal and general encomium, he finds himself forced to make quite a large number of candid concessions to its critics: "A few of his definitions must be admitted to be erroneous. . . . His definition of *Network* has been often quoted with sportive malignity, as obscuring a thing in itself very plain. . . . His introducing his own opinions, and even prejudices . . . cannot be fully defended." Nevertheless he enters a feeble defence: "Let it, however, be remembered, that this indulgence does not display itself only in sarcasm to others, but sometimes in playful allusions to . . . his own laborious task," and he quotes the definition *"Lexicographer,* a harmless drudge." Then there is the notorious "dark hints" passage, where Boswell professes it necessary to point out that, "like many other good and pious men, among whom we may place the Apostle Paul," Johnson "was sometimes overcome" by his "amorous inclinations." And he goes on and on, magnanimously extenuating Johnson's alleged frailty (for which there is not the least real evidence). Save us, indeed, from the candid friend. It seems strange that Boswell should have suppressed the innocent, and rather pitiful, story of Johnson's toying with Mrs. Desmoulins on his bed, while publishing these innuendoes of more serious lapses from continence. But the *Life,* on the whole, *is* a very strange work.

If the hypothesis is too hard to swallow that Johnson, in Boswell's *Life,* is substituting for Lord Auchinleck in more ways than one, let us abandon it, though the phenomenon is one whose existence is as well established as anything in psychology can be. The fact remains that in most readers' minds, after finishing the *Life,* there lingers around the image of Johnson an aura of unpleasant pomposity. This is not present, I think, when one finishes Hawkins's *Life,* for all that Hawkins sometimes attacks, openly and directly, actions or attitudes of Johnson that he disapproves of, and certainly not when one reads Johnson's own letters and other writings. Why is it so in Boswell? If such things as the conferring of the perpetual doctorate on Johnson—the only case in literary history of a major writer's being regularly so stigmatized—were not unconscious malice on Boswell's part, then they were certainly the result of poor judgment, or, not to put a fine point on it, obtuseness. And perhaps the whole impression of obtuseness—or "simple-mindedness"—that surrounds the Johnson of Boswell's *Life* is a reflection of the biographer rather than his subject. For—let us face it, let us clear our minds of cant—without necessarily subscribing to the whole of Macaulay's indictment of Boswell's idiocy, it must be recognized that Johnson's mind was one of the rare first-rate ones, Boswell's at best second-rate.

If someone indignantly interjects, "How then account for the tremendous popularity of Boswell's *Life?*" the regretful answer must be that it is only too readily accounted for, on the familiar ground that the good is the enemy of the best. Given the slightest excuse to avoid taking the work of a serious writer at its full seriousness, the average reader will leap at it, and get great comfort from being able to smile condescendingly with Boswell at Johnson's foibles (or with Jones at Freud's). The popularity of the cutting-down-to-size process is proved again and again in literary history; "resistance by partial incorporation," some psychologists have called it. It happened to Shakespeare, for instance, of whom several generations parroted the lament that "he wanted art," that it would have been better for him to have gone to college (like the reader) and learned to be well-bred and decently restrained in his writing, and whom later generations transmuted into the sentimental Bard of Henry Irving and the rest. The greater the writer, the more penetrating his insight into the realities of life, the more eagerly the average person will grasp at any watered-down version that some lesser man constructs and offers him as a substitute for the real thing—in one's more despondent moods, one sometimes feels that the main purpose of the academic teaching of literature has been to dilute and explain away great writers in this fashion.

My complaint against Boswell, then, is that he has only too skilfully given his public what they—and Boswell—wanted: instead of the disturbing reality that was Samuel Johnson, a cosy, "lovable," predictable, forgivable, and ultimately "safe" figure, the "dear old Doctor Johnson" of the Toby mugs, the churchwarden pipes, and the Cheshire Cheese (with none of which there is the slightest evidence that the real Johnson ever had any contact). It is no wonder that men with incisive minds from Blake to H. L. Mencken, no wonder that the brighter college students, have detested him—Boswell's Johnson, the "great Clubman" (F. R. Leavis's epithet), not the real Johnson, the "great highbrow," whom they don't know, but would respect if they did.

Johnson and Boswell

by Richard D. Altick

The admirable fullness of portraiture which is one of the book's greatest qualities was the result partly of chance and partly of Boswell's intelligence and industry. It was chance that its subject was a man who had expressed himself so copiously and quotably on so many subjects, and thus left his biographer ample materials for the delineation of his mind. It was either chance or remarkable foresight that led Boswell to keep voluminous journals long before his biographical purpose had crystallized. But it was no accident that Boswell, far from being swamped by his data, knew how to go about making a book out of them. He had a clearly formulated ideal of biography, and he saw how he could realize it. Instead of the flatness that inevitably results from use of a single vantage point, he aimed for a three-dimensional effect by, as he said, "an accumulation of intelligence from various points, by which [Johnson's] character is more fully understood and illustrated." Wherever possible, he quoted from Johnson's "own minutes [private diaries], letters, or conversation, being convinced that this mode is more lively, and will make my readers better acquainted with him, than even most of those were who actually knew him, but could know him only partially." To Johnson's own words, Boswell added the constant testimony of an eyewitness—himself—whose powers of observation and retention, cultivated over many years, were never more acute than when in Johnson's presence. The third indispensable element was the recollections of the many persons who had known Johnson before Boswell came on the scene, or who were present on occasions when Boswell was absent, or who could provide additional details of scenes that he did describe first-hand. "I will venture to say," Boswell wrote with no exaggeration whatsoever, "that [Johnson] will be seen in this work more completely than any man who has ever yet lived."

No previous biographer had been a tithe as assiduous as Boswell was in pursuit of the hitherto uncollected fact. Dr. Johnson himself had been an armchair researcher whose sources were, for the most part, limited to the books within reach. Boswell, persuaded that the authenticity and lifelikeness of a biographical portrait are directly proportional to the amount of material drawn upon and the variety of sources from which it is obtained, cast as wide a net as humanly possible. He was as expert as any modern biographer in reconstructing single episodes from several sources. His description of Johnson's conversation with George III, for example, was synthesized from information obtained from Johnson himself; from Bennet Langton, who heard Johnson tell about it one day at Sir Joshua Reynolds'; from a letter Johnson's printer-friend, William Strahan, wrote to Bishop Warburton; and from other private documents and oral reminiscences.

Eighteenth-century biographers habitually praised the accuracy of their own accounts, but few had much reason to do so. Boswell was the first to back up a claim of "scrupulous authenticity" with a record of extensive and painstaking inquiry. Johnson was eager enough to destroy a biographical myth if he could do so by use of information at hand or by simple reliance on his native shrewdness, but not if it required much exertion. Boswell on the other hand was tireless in his anxiety to establish the exact fact. He would run half over London, as he said, in order to fix a date correctly. And though many of the unkind remarks he dropped regarding the books by Sir John Hawkins and Mrs. Piozzi undoubtedly were motivated by personal jealousy— their authors were, after all, his rivals for the limelight—they also sprang from his honest intolerance of inaccuracy and bias and his desire to root out the considerable amount of legend from the Johnsonian story before it could burgeon any further. Boswell's zeal for accuracy was worth the pains it cost him. Modern scholarship, which has examined his every page with the most minute care, has found very little to correct.

If Boswell paid far more than lip service to the ideal of biographical accuracy, he also differed from his predecessors in actually avoiding the uncritically eulogistic mood which they repudiated, *pro forma,* and then adopted. "I profess," he announced, "to write, not his panegyric, which must be all praise, but his Life; which, great and good as he was, must not be supposed to be entirely perfect. To be as he was, is indeed subject of panegyric enough to any man in this state of being; but in every picture there should be shade as well as light." Hannah More, Johnson's pet in the days when she frequented London literary circles, had begged Boswell to "mitigate" some of the "asperities" of "our virtuous and most revered departed friend." Boswell retorted that "he would not cut off his claws, nor make a tiger a cat, to please anybody." His utter frankness in delineation was the main reason why the *Life of Johnson* made such a stir, and it remains today one of the book's most remarkable qualities. Once in a while, to be sure, Boswell toned down a few details, such as the bawdy talk in which Johnson occasionally indulged, especially when Garrick goaded him on. Nevertheless, the immense amount of research that has been done on Johnson's life and character has exposed no significant respect in which Boswell suppressed or modified the truth. No discreditable episode in Johnson's life, no disagreeable trait of behavior, went unmentioned in his pages.

Some of Boswell's contemporary critics maintained that in portraying Johnson's less statuesque side he had destroyed a hero. No dispassionate reader of the *Life* would agree. Johnson is not a whit less great, is, some would say, all the more admirable a man, for the way Boswell depicts him. His very weaknesses, which were serious, make his strengths heroic. Actually, what the critics complained of was Boswell's insistence on the revealing, private detail rather than what tradition considered the proper concern of biography, the great and memorable act. Johnson's greatness was so evident to him that Boswell assumed it would be obvious to everybody else. What he wanted to present to his own age, and to posterity, was not a walking set of principles but a human being: not a posed figure who could be admired from a reverential distance, but a puffing, muttering, grimacing, shambling, pocked, untidy, half-blind, rude, contentious, dogmatic, superstitious, intolerant man whom any number of people not only admired but unaffectedly loved.

Boswell chose to make Johnson come alive in his pages through the use of lavish but controlled detail: each detail significant and revealing in itself, and the total cumulative in effect. He mingled the material of Flemish realism (an analogy he himself was aware of) with what was to be, a century later, the method of French neoimpressionism. His raw material was the individually minute data of the senses— sharply observed particulars of personal appearance, dress, conduct, peculiarities of speech, locale. From these thousands upon thousands of small details, carefully arranged on the broadest canvas a biographer had ever commanded, Boswell produced the vivid portrait—or whole set of portraits—which makes the *Life of Johnson* a masterpiece.

Boswell gives us incomparably more to see and hear than any preceding biographer. When Boswell and Johnson encounter Johnson's old schoolfellow, Oliver Edwards, we recognize the prosy bore as the very model of the person most of us discover our own schoolmates to have turned into at the interval of several decades; Boswell places him, once and for all, by a sensitive reproduction of his conversation. When an episode occurs in a street, the street is duly named, so that everyone who knows eighteenth-century London can envision the exact locale—

buildings, sidewalks, gutters, street cries, smells, and all. When a boy in a water taxi pleases Johnson and Boswell, they reward him not with a (generalized) coin but with a (specified) shilling. Above all, the manifold physical eccentricities and compulsive mannerisms of Johnson are described with an unflinching particularity that has engraved them in the memory of readers for over a century and a half.

How novel this kind of reporting was in Boswell's time is suggested by the frequency with which he calls attention to it in order to defend it. He is almost oppressively conscious of the disapproval that hangs in the critical atmosphere of an age still anxious for the avoidance of "vulgar" particulars and the achievement of general (and therefore philosophically significant) effects. "I cannot," he says on one occasion, "allow any fragment whatever that floats in my memory concerning the great subject of this work to be lost. Though a small particular may appear trifling to some, it will be relished by others; while every little spark adds something to the general blaze: and to please the true, candid, warm admirers of Johnson, and in any degree increase the splendour of his reputation, I bid defiance to the shafts of ridicule, or even of malignity." With which magnificent gesture—Boswell against the critics—he proceeds to increase the splendor of Johnson's reputation by a rather circuitous route; for the following scene is in the country, where Johnson, on a visit to a friend, is trying to clear an artificial waterfall of accumulated débris. "He worked till he was quite out of breath; and having found a large dead cat so heavy that he could not move it after several efforts, 'Come,' he said, (throwing down the pole,) *'you* shall take it now;' which I accordingly did, and being a fresh man, soon made the cat tumble over the cascade." Johnson the great moralist and man of letters is not immediately evident in the scene, but Johnson sweating to dislodge a dead cat from a waterfall nevertheless is the man who wrote *The Vanity of Human Wishes,* and Boswell's great achievement is that he leaves no doubt in our minds of their identity.

In addition to being a supremely gifted portraitist, Boswell was a born dramatist. The time being what it was, an age of intense sociability; and Dr. Johnson being the man he was, loving nothing more than to sit with chosen company and talk away the evening; and Boswell being the man *he* was, a constant playgoer whenever he was in London, it was almost inevitable that the book should have been deliberately planned as a series of scenes. Boswell had the priceless advantage of having been present on hundreds of occasions when there was free discourse between Johnson and one or more interlocutors, many of them men of high intellect whose conversation put him on his mettle, others being fools, lightning rods to bring down the bolt of his devastating wit. Boswell had made full records of those scenes, as well as of the equally numerous times when he and Johnson were alone, in Johnson's rooms or strolling through the streets. The book's strong dramatic element, in fact, was already in existence when Boswell transferred the Johnsonian passages of his private journals into the manuscript of the *Life*. It took no such labor as is usually the lot of the modern biographer who seeks to dramatize his pages by a painstaking assembling of facts, but the effect of this spontaneous staging, so to speak, is all that the serious biographical artist could wish. The man who is the subject of the biography is constantly placed in a social milieu. He is surrounded by others, and their interaction, as mirrored by their conversation, provides an unexcelled immediacy of impression, both of him and of them.

Few of the scenes in Boswell mark crucial episodes in Johnson's life. But Boswell was always less interested in the strictly narrative side of his study than in portraiture. And Johnson could be most accurately and extensively portrayed in the midst of a company, talking and, in the process, unfolding his mind. Hence Boswell uses scenes (many of which he adroitly stage-managed with such results in view) for the sake of character revelation. With the assistance of the people present at the Mitre or the Thrales', Johnson from 1763 to 1784 literally talked himself into his own biography. The constant conversation in the book, whether Johnson is responding to Boswell's inquisitive prodding as they sit alone or is participating in a free-for-all in a drawing room whose air is pungent with dispute, accounts for as much of its vividness as does Boswell's evocation of physical appearances. Though the immediate interest is in the juxtaposition of personalities and the collision of ideas, in the end the effect is what Boswell strove for above all: the reader has learned to know a man.

Along with its gigantic merits, however, the *Life of Johnson* has certain qualities which, to a modern critic, make it less than a perfect example of biographical art. Nowadays, thanks to the recovery of the fabled Malahide papers, it is possible to study how Boswell transformed his jottings into a printed masterpiece; and as a result nobody subscribes to Macaulay's once widely shared view of him as a sottish fool who somehow blundered into writing a work of genius. Boswell was an artist, and he knew what he was doing. But his artistry was, in some ways, limited, though less by any personal deficiencies than by the conditions of his age and the special circumstances that led to the writing of his book.

Those circumstances, to be sure, probably enabled Boswell to make a better book than he otherwise could have, because they were such as to encourage the exercise of his special talents. He was an artist who was best fitted for small-scale operations: the management of individual scenes, the writing of descriptive passages—detail work rather than architecture. His genius was less adapted to solving the larger problems of structure and proportion. The materials he possessed and the particular nature of his own experience with the subject of his biography required, for best utilization, precisely the sort of creative gift he brought to the task.

Nevertheless, it must not be forgotten that the *Life of Johnson* was written principally from notes made during, and dealing with, a relatively limited period of its subject's life. Johnson was born in 1709; Boswell first met him in 1763, when he was a few months short of fifty-four; Johnson died in 1784. Yet the whole first half of Johnson's life is condensed into barely one tenth of the biography, whereas the final eight years of Johnson's seventy-five spread over no less than half the pages.

The heart of the book is Boswell's first-hand reporting of Johnson as he knew him. But not only were the two men acquainted only in the last twenty-one years (considerably less than a third) of Johnson's life; during that period they were in the same vicinity for a total of less than two years and two months, and when they were apart, there were long lapses in their correspondence. So, even despite Boswell's praiseworthy attempts to make up for this limitation by tapping the memories of Johnson's other associates, by far the major part of the *Life* is devoted to a small group of segments of Johnson's whole career. The book is decidedly out of balance. It is essentially a report of the older Johnson as Boswell happened, from time to time, to see him.

Again, the fact that Boswell's journals formed the basis of the book from 1763 onward had, along with its manifest advantages, certain disadvantages. As we have noted, the nature of those memoranda, with their great emphasis on conversation, virtually required that the biography be cast as a long series of scenes. Even after making all allowance for Boswell's efforts to fill in from other sources, it remains true that the majority of these scenes are presented from the viewpoint of James Boswell. Boswell may not unfairly dominate the picture, as he used to be accused of doing, but his presence is obvious enough. Johnson is seen chiefly through his eyes and ears, simply because it was through them that the raw stuff of the *Life,* the contents of the journals, was acquired.

Though one would not for a moment wish Boswell to have sacrificed his scenes, they did put him in something of a straitjacket. Down to 1763, his story of Johnson's career had to be synthesized from second-hand materials. Boswell was able, therefore, to construct a smooth-running narrative, interweaving the events of Johnson's private life with the progress of his literary career. He was free to rearrange and organize, in any way that seemed proper, the assorted facts he had gathered from Johnson and others. It was a fairly simple job of retrospective reconstruction. But abruptly, in 1763, the whole character of his source material changed. Henceforth the journals would, in effect, determine the form the biography would take. Extended scenes, hitherto very infrequent, would predominate, and the summary narrative would accordingly diminish in importance. When Boswell takes up his role of first-hand observer, the whole technique of the book shifts from the essentially narrative to the dramatic mode.

The primacy of the journals as source material had one other regrettable effect. Their structure was, of course, rigidly chronological; they were a day-by-day record of Boswell's life. An orderly narrative of events is desirable, indeed requisite, in any biography. But where the calendar arbitrarily dictates the biographer's direction, there is bound to be a miscellaneity of effect. The straighter the chronological course, the more rapid and abrupt the tacking from one subject to another. Within the framework of a single year, and profusely interspersed with one another, occur discussions of Johnson's external life, literary activities, domestic events, travels, emotional and intellectual tendencies as expressed by letters and conversations dating from that year, and whatever else the records assign to the stated period.

The predominance of scenes in Boswell intensifies this disorderliness. Reporting each conversation as it happened, Boswell had little choice but to follow the stream of discussion wherever it led—to the views of Johnson and others on religion, philosophy, politics, on the vagaries of human behavior, on individual personalities, on Johnson himself. Because Johnson returned to the same subject—death or his love of good eating—and gave instances of his kindness or his irritability a score of times over the years Boswell records, the separate bits of data normally are scattered throughout the book. Admittedly, life is like that, and the fidelity with which Boswell reflects the shifting interests of Johnson and his conversational circle contributes its substantial share to the pervasive verisimilitude. But one of the purposes of art is to reduce the disorderliness of life.

The enduring fame of the *Life of Johnson* as a browsing book, to be opened at random, is itself an indication of its lack of taut organization. The very circumstance that Boswell chose to transfer his journal entries more or less *en bloc* to the book, concentrating his energies on the touching up of detail rather than on a sweeping reordering and reassessment of material, suggests a deficiency in his equipment as an artist. He was somewhat lacking in a sense of relative values. His anxiety not to neglect the apparently trivial which actually had relevance led him to include, also, the genuinely trivial. Between the familiar peaks of interest, from which no one would wish a single word to be subtracted, are long level stretches which Boswell could have drastically shortened. He was so engrossed in his wealth of source material that he could not see it in perspective. A greater artist would have taken more care to distinguish among the various gradations of importance and interest that his raw data contained.

PART THREE

View Points

A. Edward Newton

The great scholar Jowett confessed that he had read the book fifty times. Carlyle said, "Boswell has given more pleasure than any other man of this time, and perhaps, two or three excepted, has done the world greater service." Lowell refers to the "Life" as a perfect granary of discussion and conversation. Leslie Stephen says that his fondness for reading began and would end with Boswell's "Life of Johnson." Robert Louis Stevenson wrote: "I am taking a little of Boswell daily by way of a Bible. I mean to read him now until the day I die." It is one of the few classics which is not merely talked about and taken as read, but is constantly being read; and I love to think that perhaps not a day goes by when some one, somewhere, does not open the book for the first time and become a confirmed Boswellian.

From "James Boswell—His Book" by A. Edward Newton, in The Amenities of Book-Collecting and Kindred Affections (Boston: The Atlantic Monthly Press, 1918), p. 185. Reprinted by permission of Atlantic—Little, Brown and Co.

Chauncey Brewster Tinker

There is a certain kind of reader who vexes himself and teases the critic with the question whether the author of a great classic really put into it all that an enthusiastic reader asserts that he finds. Is it a conscious art, or has all the greatness, all the subtlety and meaning of it, been thrust upon it by the critic? A suspicious reader can usually be set right by passages in which the author himself has spoken of his art. A critic is as little likely to see more than he was intended to see as a stream is likely to rise above its source. If anybody doubts whether Boswell meant to produce the effects for which he is famous, let him gather up everything that the man said about his art, about Johnson's theory of biography, and, above all, everything that he said

From "The Magnum Opus" by Chauncey Brewster Tinker, in Young Boswell (London: G. P. Putnam's Sons; Boston: The Atlantic Monthly Press, 1922), pp. 220-24. Reprinted by permission of Atlantic—Little, Brown and Co.

about his own books, and he will convince himself that Boswell's effects were all calculated.

George Gordon

It is late in the day to be advancing the merits of Boswell's *Life of Johnson*, and pressing its claims as a companionable book. Probably no English publication of the last hundred and thirty years has made more friends or kept them longer. Its votaries are of all ages and both sexes, and their number, which has always been large, seems to be constantly increasing. It has increased very notably in the last twenty years. There is something in the character of Johnson, and in Boswell's portrait of him, which evidently appeals with peculiar force to the age in which we are now living. I suppose it is partly a great weariness of make-believe that has directed so many eyes upon him, and is now replacing the spent wind of Victorian idealism with his Georgian robustness and his majestic common sense. We are a somewhat disillusioned generation. The rainbow promises of our fathers have not been kept, and we turn with relief to this sworn enemy of cant, who never pretended even to himself that life can yield more than we are willing to put into it, or that Utopia can be reached by exhalations of the breath. "When a butcher tells you that his heart bleeds for his country, he has in fact no uneasy feeling."

We are conscious also—which is another reason—standing among the scientific wonders of our day, that while we have gained in power, we have lost in art, and most notably, perhaps, in the chief art of all. Human power is enormous, but in the chaos of new contrivances we have somehow contrived to lose the art of living. Now this, as it happens, is precisely the art which the eighteenth century and Johnson have to teach us. It is an art of dignity, simplicity, and quiet, and by a kind of homing instinct we are returning to it.

I said that the devotees of Boswell's *Johnson* are of both sexes and of all ages. Yet it is a man's book, and its talk is men's talk. It is humorous, but also profoundly rational, and hardly anything in it could have been said by a woman, or, for that matter, by a lover or a child. Children appear only as a topic, to have their education settled, and women, for the most part, as a social problem. The affair of love, on the rare occasions when it is mentioned, is treated either as a theme for poets, or as an occasion for prudence. To some extent this

From "Boswell's Life of Johnson" by George Gordon, in Companionable Books: Series I. (London: Chatto and Windus Ltd., 1927), pp. 45–48. Reprinted by permission of Dr. George Gordon and Chatto and Windus Ltd.

is a result of the Boswellian method. The Johnson who played the elephant with the little Thrales,—who, if he had had "no duties, and no reference to futurity," would have spent his life "in driving briskly in a post-chaise with a pretty woman,"—the Johnson who had "more fun, and comical humour, and love of nonsense about him" than almost anybody Fanny Burney ever saw, is sparsely represented in Boswell's pages. No man more than Johnson enjoyed the society of the tea-table, or set a higher value on the company of elegant, sensible, and vivacious women. The happiest time of his life, he told Mrs. Thrale, was when he spent "one whole evening" talking with Molly Aston—"the loveliest creature I ever saw," and a wit and a scholar besides. "That, indeed, was not happiness, it was rapture; but the thoughts of it sweetened the whole year." This side of Johnson was not to be ignored, but it made Boswell jealous and uneasy. He had neither eyes nor ears for the ladies when Johnson was in the room, and seems almost to have grudged his hero's gallantry because it offered so little to the reporter. This was not, I must add, because Boswell was unsusceptible of female charm. "I got into a fly at Buckden," he writes to a friend, "and had a very good journey. An agreeable young widow nursed me and supported my lame foot on her knee. Am I not fortunate in having something about me that interests most people at first sight in my favour?" This was in a fly, when he was off duty. In Johnson's company his business was with Johnson, and Johnson, he thought, was more Johnsonian among men. I have no doubt that he was right. But there was a great deal of Johnson.

The book, then, is masculine, though not forbiddingly so. I first met it as a schoolboy, and remember still the almost magical impression of it. Here, by turning a few pages, I found myself admitted not only to a larger world, but actually to a Club, and, as I was to verify later, the best Club in literature. My case is not uncommon. The late Sir Leslie Stephen declared, at the close of a life devoted to authorship and letters, that his enjoyment of books had begun and ended with Boswell's *Life of Johnson*.

Though it suits all ages, it is a book, I fancy, best appreciated in the middle years, and by those who have had to fight for their experience, who have not found life easy, and who are still in the battle. Intelligence is not enough, even superior intelligence, as Macaulay proved. No admirer of this book has more disastrously misunderstood it. To understand Johnson it is necessary to have lived and to have thought about life, for life was his trade.

William Lyon Phelps

Coming back for the last time at present to Dr. Johnson, I hope it is not an impertinence to my readers to suggest that they provide themselves with a good copy of Boswell's *Life of Johnson*. And for two reasons. It is the best bed book I know. I believe one could read it through three times a year with unflagging interest. Second, this book, unlike books designed for children, and books that may be read with delight by children although never intended for them (in this second class are *Gulliver's Travels, Robinson Crusoe, Pilgrim's Progress,* and the *Old Testament*) Boswell's *Life of Johnson* is intended *exclusively for adult readers.* In times when so much is done by publishers, motion picture directors, and radio for infants and for men and women with infantile minds, it is refreshing to have at least one great classic that can be appreciated only by men and women who are mentally mature.

From "Esquire's Five-Minute Shelf" Esquire, XIV (September 1940), 170–71. Reprinted by permission of Esquire Magazine.

Chronology of Important Dates

Johnson		Boswell
1709	Sept. 7 (old style), Samuel Johnson born in Lichfield, Staffordshire, son of a local bookseller.	
1717–26	Educated at Lichfield grammar school.	
1728	Late October, enrolls at Pembroke College, Oxford and remains until mid-December 1729.	
1731	Father dies.	
1735	Marries "Tetty"—a widow 45 years old—and opens school at Edial, near Lichfield.	
1737	Comes to London seeking employment.	
1738–1746	Working for *Gentleman's Magazine,* and as a Grub Street writer.	
	1740	Oct. 29 (new style), James Boswell born in Edinburgh, Scotland, son of Alexander Boswell, Lord of Auchinleck, a Lord of Session.
1746–55	Working on *Dictionary.*	
1749	January, *The Vanity of Human Wishes.* February, *Irene* produced at Drury Lane Theatre.	

1750–52	*The Rambler.*		
1752	March, his wife dies.		
		1753–58	Enrolled at University of Edinburgh.
1755	*Dictionary* published, establishes his fame.		
1758–60	*The Idler.*		
1759	*Rasselas.*	1759	Autumn to February 1760 at University of Glasgow.
		1760	Becomes Roman Catholic (for about a month). In March runs off to London. Father comes down to get him. End of May, both return to Scotland.
1762	July, given a pension by King George III.	1762	September, begins keeping a full journal. November, goes to London.
1763	May 16, meets Boswell.	1763	Meets Johnson in Tom Davies' back parlor. Early August, goes to Continent to study law.
		1763–64	Attending University of Utrecht in Netherlands.
1764	Spring, formation of The Literary Club.	1764–65	Continental tour, meets Rousseau and Voltaire.
1765	Edition of Shakespeare. January, meets the Henry Thrales. From 1766 to about 1782 lives with them part of the time.	1765	October, goes to Corsica and meets General Paoli.
		1766	February, returns to London, rushes to see Johnson, then returns to Edinburgh and is admitted to the bar.
		1768	Publishes *Account of Corsica.* In spring, visits Johnson in Oxford.

1769   Autumn, sees Johnson in London. Nov. 25, marries cousin Margaret Montgomerie in Scotland.

1770   *The False Alarm* (political pamphlet).

1772   In London in the spring

1773   August to November with Boswell in Scotland.

1773   In London in spring. Elected to The Club with Johnson as sponsor. Autumn tour to the Hebrides with Johnson. Sees Johnson in London in the spring of 1775, 1776, 1778, 1779, 1781, 1783 and summer of 1784. Meets him in Ashbourne in autumn of 1777.

1774   Tours Wales with Thrales.

1775   *Journey to the Western Islands of Scotland* published. Receives honorary D.C.L. from Oxford. Goes to Paris with Thrales in autumn.

1777   Agrees to write prefaces for edition of British poets.

1777   October, begins to write *Hypochondriack* essays for *London Magazine.* Continues through August 1783.

1779   First four volumes of *Lives of the Poets* published.

1781   Henry Thrale dies. Last volumes of *Lives* published.

1783   June, suffers a stroke, with temporary loss of speech.

1784   July, Mrs. Thrale marries Italian musician Gabriel Piozzi. December 13, dies and on the 20th buried in Westminster Abbey.

1785	September, publishes *Tour to the Hebrides*.
1786	Admitted to the English bar. Moves family to London. Begins serious work on the *Life*.
1789	Wife dies.
1791	May 16, *Life of Johnson* published.
1793	2nd edition of *Life*.
1795	May 19, dies and is buried at Auchinleck.

Notes on the Editor and Contributors

PAUL K. ALKON, a member of the Department of English at the University of Maryland, is the author of *Samuel Johnson and Moral Discipline*.

RICHARD D. ALTICK is Professor of English, Ohio State University. Among his published works are *The Scholar Adventurers, Lives and Letters*, and *The Art of Literary Research*.

JAMES L. CLIFFORD is Trent Professor of English Emeritus, Columbia University. He is the author of *Hester Lynch Piozzi (Mrs. Thrale), Young Sam Johnson*, and *From Puzzles to Portraits: Problems of a Literary Biographer*.

GEORGE GORDON was at various times President of Magdalen College, Professor of Poetry, and Vice Chancellor of Oxford University. Among his critical works are *The Discipline of Letters* and *Companionable Books*.

A. S. F. GOW is a Fellow of Trinity College, Cambridge, and a well-known classical scholar. He has published widely on Theocritus and the bucolic poets, and is the author of a study of A. E. Housman.

DONALD J. GREENE, Bing Professor of English, University of Southern California, is the author of *The Politics of Samuel Johnson* and *The Age of Exuberance: Backgrounds to Eighteenth-Century English Literature*, and has contributed widely to a new critical evaluation of the period.

LIEUT.-COL. RALPH HEYWARD ISHAM, collector and raconteur, was influential in bringing together in one place all of the Boswell papers. His great collection is now largely at Yale University.

JOSEPH WOOD KRUTCH, formerly Brander Matthews Professor of Dramatic Literature, Columbia University, died in Tucson, Arizona in May 1970. He had written extensively on literature and on Nature. Among his books are *Comedy and Conscience after the Restoration, The Modern Temper*, and *Samuel Johnson*.

A. EDWARD NEWTON was a famous collector who lived at "Oak Knoll," Berwyn, Pennsylvania. Some of his essays, which first appeared in *Atlantic Monthly*, may be found in *The Amenities of Book-Collecting* and *A Magnificent Farce*

SIR HAROLD NICOLSON, diplomat and author, wrote numerous biographies—of Verlaine, Tennyson, Benjamin Constant, and others—and a short history of the genre. Since his death, three volumes of selections from his diary and letters have been published by his son.

WILLIAM LYON PHELPS was for many years a popular teacher at Yale University and a well-known lecturer. He was the author of *The Advance of the English Novel* and *The Advance of English Poetry*.

FREDERICK A. POTTLE, Stirling Professor of English Emeritus, Yale University, is the chief living authority on Boswell. He has edited many volumes of Boswell's journals and has published the first installment of his biography. Among Professor Pottle's other works are *The Literary Career of James Boswell* and *The Idiom of Poetry*.

GEOFFREY SCOTT, a brilliant young scholar, who died in 1929, at the age of 44, was editor of the first six volumes of the *Private Papers of James Boswell from Malahide Castle*, and is the author of *The Architecture of Humanism* and *The Portrait of Zélide*.

FRANK TAYLOR is Keeper of Manuscripts of the John Rylands Library in Manchester, England. Most of his printed work deals with the library's collections and includes calendars of English, Latin, French, German and Dutch manuscripts and studies of individual manuscripts (fifteenth–eighteenth centuries).

CHAUNCEY BREWSTER TINKER, whose course at Yale on "The Age of Johnson" was immensely popular, inspired many students who have been active in the revival of interest in the eighteenth century. Editor of *The Letters of James Boswell*, he was the author of *Nature's Simple Plan* and *Painter and Poet*.

MARK VAN DOREN, best known as a poet, was before his retirement a Professor of English at Columbia University. He now lives at Falls Village, Connecticut. He is the author of *John Dryden, Don Quixote's Profession, The Happy Critic*, and numerous volumes of poetry.

MARSHALL WAINGROW, Professor of English, Claremont Graduate School, is now engaged in editing the original manuscript of Boswell's *Life of Johnson*.

Selected Bibliography

Editions

The standard edition of the *Life of Johnson* is that edited by G. B. Hill, revised and enlarged by L. F. Powell; Vols. I-IV (1934); Vols. V-VI (1950, revised 1964). A revised edition of I-IV is expected soon.

The best of the inexpensive modern editions are the Oxford Standard Edition, introduction by C. B. Tinker (1933; paperbound 1960); and the Everyman's Library, introduction by S. C. Roberts (2 vols., 1949). Both have useful indexes.

Of the many abridged paperback editions, three may be mentioned: that edited by Anne and Irvin Ehrenpreis (Washington Square Press, 1965); by Frank Brady (New American Library, Signet Classics, 1968); and Robert Hunting (Bantam Books, 1969).

Commentary

In addition to those works cited in the notes to the Introduction to this volume and those from which excerpts are included in the main body of the text, the following may be recommended for serious students wishing to do further research on the topic:

Amory, Hugh, "Boswell in Search of the Intentional Fallacy," *Bulletin of the New York Public Library,* LXXIII (January 1969), 24–39.

Baldwin, Louis, "The Conversation in Boswell's *Life of Johnson,*" JEGP, LI (October 1952), 492–506.

Bronson, Bertrand H., "Boswell's Boswell," *Johnson Agonistes.* Cambridge University Press, 1946, pp. 53–99.

Brown, Anthony E., "Boswellian Studies: a Bibliography," *Cairo Studies in English, 1964,* ed. Magdi Wahba. Cairo, UAR, 1966, pp. 1–75.

Butt, John, "James Boswell," *Biography in the Hands of Walton, Johnson and Boswell* (Ewing Lectures). Los Angeles: Univ. of California, 1966, pp. 33–48.

Carver, George, "Boswell and the 'Johnson,'" *Alms for Oblivion: Books, Men and Biography.* Milwaukee, Wis.: Bruce Publishing Co., 1946, pp. 160–69.

Chapman, R. W., "Boswell's Revises," *Johnson & Boswell Revised by Themselves and Others.* Oxford: Clarendon Press, 1928, pp. 21–50.

————, "The Making of the *Life of Johnson*," *Johnsonian and Other Essays and Reviews*. Oxford: Clarendon Press, 1953, pp. 20–36.

Collins, P. A. W., *James Boswell*, Writers and Their Work. London: Longmans, Green & Co. Ltd., 1956.

Davis, Bertram H., *Johnson Before Boswell: a Study of Sir John Hawkins' "Life of Samuel Johnson."* New Haven: Yale Univ. Press, 1960.

Drinkwater, John, "Johnson and Boswell," *The Muse in Council.* London: Sidgwick and Jackson, 1925, pp. 218–24.

Edel, Leon, *Literary Biography.* Toronto: Univ. of Toronto Press, 1957, pp. 13–20.

Fussell, Paul, "The Memorable Scenes of Mr. Boswell," *Encounter.* XXVIII (May 1967), 70–77.

Hart, Edward, "The Contributions of John Nichols to Boswell's *Life of Johnson*," *PMLA*, LXVII (June 1952), 391–410.

Jack, Ian, "Two Biographers: Lockhart and Boswell," *Johnson, Boswell and Their Circle: Essays Presented to L. F. Powell.* Oxford: Clarendon Press, 1965, pp. 268–85.

Longaker, John Mark, "Boswell's *Life of Johnson*," *English Biography in the Eighteenth Century.* Philadelphia: Univ. of Pennsylvania Press, 1931, pp. 407–76.

Lustig, Irma S., "Boswell's Literary Criticism in *The Life of Johnson*," *Studies in English Literature.* (Rice University), VI (Summer 1966), 529–42.

Morgan, Lee, "Boswell's Portrait of Goldsmith," *Studies in Honor of John C. Hodges and Alwin Thaler*, ed. R. B. Davis and J. L. Lievsay. Knoxville: Univ. of Tennessee Press, 1961, pp. 67–76.

Pearson, Hesketh, "Truth in Biography," *Ventillations: Being Biographical Asides.* Philadelphia and London: J. B. Lippincott Co., 1930, pp. 11–18.

Pottle, Frederick A., *The Literary Career of James Boswell.* Oxford: Clarendon Press, 1929).

Powell, L. F., "The Revision of Dr. Birkbeck Hill's Boswell," *Johnson & Boswell Revised by Themselves and Others.* Oxford: Clarendon Press, 1928, pp. 53–66.

Rader, Ralph W., "Literary Form in Factual Narrative: the Example of Boswell's *Johnson*," *Essays in Eighteenth-Century Biography*, ed. Philip B. Daghlian. Bloomington: Indiana Univ. Press, 1968, pp. 3–42.

Reid, B. L., "Johnson's Life of Boswell," *Kenyon Review*, XVIII (Autumn 1956), pp. 546–75; reprinted in *The Long Boy.* Athens, Ga.: Univ. of Georgia Press, 1969, pp. 1–30.

Stauffer, Donald A., "The Great Names," *The Art of Biography in Eighteenth Century England.* Princeton: Princeton University Press, 1941, pp. 411–55.

Wilson, F. P., "Table Talk," *Huntington Library Quarterly*, IV (October 1940), 27–46.

Wimsatt, William K., Jr., "James Boswell: the Man and the Journal," *Yale Review*, XLIX (Autumn 1959), 80–92; reprinted (revised) as "The Fact Imagined: James Boswell," in *Hateful Contraries.* Lexington: Univ. of Kentucky Press, 1965, pp. 165–83.